Corralling The Colorado

The First Fifty Years
of the
Lower Colorado River Authority

JAMES H. (JIMMY) BANKS
and
JOHN E. BABCOCK

EAKIN PRESS ★ Austin, Texas

ISBN 0-89015-662-X

Library of Congress Cataloging-in-Publication Data

Banks, James H.
   Corralling the Colorado: the first fifty years of Lower Colorado River Authority / by
James H. (Jimmy) Banks and John E. Babcock.
        p.   cm.
   Bibliography: p.
   Includes index.
   ISBN 0-89015-662-X : $19.95
      1. Lower Colorado River Authority — History. 2. Colorado River Region (Colo. –
Mex.) — History.   I. Babcock, John E.   II. Title.
TC425.C6B36   1988   88-16036
                                                                                          CIP

*To the men and women —
past, present, and future —
of the Lower Colorado River Authority*

Cooperatives Served by LCRA

Bandera Electric Cooperative, Inc. (Bandera)
Bluebonnet Electric Cooperative, Inc. (Giddings)
Central Texas Electric Cooperative, Inc. (Fredericksbur
DeWitt County Electric Cooperative, Inc. (Cuero)
Fayette Electric Cooperative, Inc. (La Grange)
Guadalupe Valley Electric Cooperative, Inc. (Gonzales
Hamilton County Electric Cooperative, Inc. (Hamilton)
Kimble Electric Cooperative, Inc. (Junction)
McCulloch Electric Cooperative, Inc. (Brady)
Pedernales Electric Cooperative, Inc. (Johnson City)
San Bernard Electric Cooperative, Inc. (Bellville)

Cities Served by LCRA

Bastrop	Gonzales	Moulton
Bellville	Hallettsville	New Braunfels
Boerne	Hempstead	San Marcos
Brenham	Ingram	San Saba
Burnet	Kerrville	Schulenburg
Cuero	La Grange	Sequin
Flatonia	Lampasas	Shiner
Fredericksburg	Lexington	Smithville
Georgetown	Llano	Waelder
Giddings	Lockhart	Weimar
Goldthwaite	Luling	Yoakum
	Mason	

Contents

Preface		vii
1.	Mainly in the Plains	1
2.	Dirty Tricks	10
3.	A Power Struggle	18
4.	The Pyramid Falls	27
5.	A Fair Trade	39
6.	Big Bargain	51
7.	"Fifty Hundred Thousand"	64
8.	Calling a Bluff	74
9.	A "Hole" in the Dam	88
10.	The Biggest Catch	99
11.	Power to the People	107
12.	High Time	119
13.	Help Wanted	126
14.	The Seguin Song	136
15.	A Model of Efficiency	143
16.	The Selling Circus	150
17.	On the Warpath	159
18.	Missions of Mercy	170
19.	Getting Steamed Up	184
20.	Gas Attacks	194
21.	The Peace Maker	212
22.	A New Beginning	224
Appendix		231
	A. Dam Specifications	231
	B. LCRA Board of Directors	233
Bibliography		239
Photo Section		243
Index		257

Preface

Many years ago, Byron Utecht, an Austin correspondent for the *Fort Worth Star-Telegram,* told a colleague that he planned to write a history of the world in 1,000 words. "How can you possibly write a history of the world in just one thousand words?" he was asked. "Well," he replied, "some of the details will have to be omitted."

The history of the Lower Colorado River Authority's first fifty years is so filled with dramatic incidents, political intrigue, legal battles, and engineering excellence that we have had to omit some of the details in this account. Our version was not intended to be a textbook-type history; rather, it attempts to tell the exciting story of the fascinating people who made this agency one of the most successful in the country. It deals with their foibles, problems, jealousies, ambitions, and personality conflicts as well as with their outstanding achievements — and with how these remarkable individuals overcame their own problems and weaknesses to corral the mighty, once vicious Colorado River.

This story was initiated by John E. Babcock, a longtime friend of mine who worked thirty-six years for the LCRA before retiring in February of 1983. After John died of a heart attack in March of 1984, I was asked to complete the project.

The cooperation I have encountered from everyone connected with the LCRA has been remarkable. I feel certain that John received from them that same willingness to help. I will not attempt to list names, for fear of inadvertent omissions, especially among some of those whom he must have interviewed. However, I do know that we both relied heavily on the remarkable memory and keen insights of District Judge Thomas C. Ferguson, a member of the original LCRA Board of Directors, and we are especially grateful for his cooperation. We also are heavily indebted not only to the

LCRA staff but also to the employees of the Austin-Travis County Historical Collection in the Austin Public Library, the Lyndon B. Johnson Library, and the Barker Texas History Center of The University of Texas.

To all those who helped both of us in this endeavor goes our deepest appreciation.

JIMMY BANKS

1

Mainly in
the Plains

It all started with a river that was beautiful and usually placid — deceptively docile most of the time but noted for flying occasionally into ferocious, murderous rages. The Colorado never was one of your "run-of-the-mill" streams. She had a personality all her own. For many years, she played the role of an attractive but fickle temptress who taunted — and even seemed to dare — the steadily encroaching tide of civilization to try taming her. She had a knack for switching suddenly, without warning, from a stream that sustained life into one that crushed it unmercifully.

Almost from the moment she was discovered, the Colorado was bathed in controversy, and there appears to be little chance that she will ever be entirely free of it. She has made waves in law, high finance, engineering, construction, government philosophy, utility regulation, and politics. She has been responsible for the making — and the losing — of fortunes in those fields by many of those who courted her.

She even provided an important stepping stone for one of her most ardent suitors, Lyndon Baines Johnson, in his long, successful quest for the presidency of the United States.

The Colorado made her own bed, at first. She gave aid and

comfort to the people along her banks, luring them to her side then just as she does today. But periodically she would switch from one bed to another, throwing her enormous weight around indiscriminately during the process. She slammed cows and horses against houses, then hurled the houses against bridges — all the way from West Texas to the Gulf of Mexico. Her floodwaters devastated not only homes but businesses and schools and farms and everything else that happened to get in her way.

This is the dramatic story of the men and women who finally brought this magnificent river under control, harnessing her tremendous power and using it to light up a region larger than some states. The Lower Colorado River Authority, which began its operations on February 9, 1935, quickly became a focal point in the nationwide battle between forces espousing public, government-sponsored power sources and those which wanted to leave the electricity business solely in the hands of privately owned companies.

The people most directly affected, those living within the forty-one counties now served by LCRA power plants, cared little about the philosophical battle raging in Austin and Washington. But as soon as they learned what electricity could do, they demanded the new way of life it would bring; it would be a slow process that, in many cases, did not hit home until their sons returned from World War II.

Meanwhile, the river just kept rolling along — for nearly 600 miles, from Dawson County near the New Mexico border to Matagorda Bay on the Gulf of Mexico. Its length and its drainage area of 39,900 square miles makes it, according to the *Texas Almanac*, the largest river lying wholly within the state of Texas. The Colorado's runoff reaches 2,000,000 acre-feet of water annually, although it begins in a semiarid tableland about 3,000 to 4,000 feet in elevation, where the annual rainfall averages only about five to fifteen inches.

While most major rivers in the western part of the United States are fed by the springtime melting of deep snows in their upper reaches, the Colorado's primary source of water is rainfall. Ironically, its rain falls mainly in the plains — the coastal plains which constitute the lower 140 miles of the river basin. The normal annual rainfall reaches about forty inches at its mouth, but approximately 11,000 square miles of the Colorado's watershed is listed by the U.S. Geological Survey as "non-contributing." Much of that area is in the High Plains region, where the river stubbornly tries to carve out a foothold before moving southwestward into the Edwards Plateau.

When it reaches the fabled Texas Hill Country, which ranges in elevation from 500 to 2,000 feet and is marked by steep slopes with shallow, rocky soils, the Colorado begins to really get down to business. Brief periods of intense rainfall in this area have triggered many floods, with most of them reinforced by waters from such tributaries as several prongs of the Concho River, Pecan Bayou (which the *Texas Almanac* describes as the farthest western "bayou" in the United States), the Llano, San Saba, and Pedernales rivers, and many small, spring-fed creeks.

When the Colorado finally escapes from this hilly terrain, it slows down and follows a meandering, snakelike course across the gentle, nearly flat coastal plains. It drops only 450 feet during the last 290 miles of its picturesque journey to the Gulf. This lower half of the Colorado was the area that enticed men, even back during the seventeenth century, to try using it for transportation. Time and time again, it thwarted their best navigational efforts.

The possibility of using the river for transportation posed only one of its major challenges. The Colorado repeatedly lashed its downstream admirers with disastrous floods; twenty-two of them, which inflicted hundreds of millions of dollars in damages and caused countless casualties, were recorded between 1843 and 1938. Hopes for storing these floodwaters and preserving them for beneficial agricultural use eventually submerged the transportation goal, especially after the railroads reached Austin in 1871.

But men still fought, for many years — with rifles, dynamite, and a wide assortment of "dirty tricks" — over the water they needed for irrigation, especially for the huge rice crops in the Lower Colorado Valley. Before the LCRA was created and built its chain of dams, the supply always seemed to be either too little or too abundant.

The political climate was equally erratic. Public sentiment for harnessing the river rose and fell almost with the level of the stream, soaring with every flood and then quickly subsiding. Ultimately, it took some sharp dealing by a much-maligned husband-and-wife team of Texas governors, James E. (Pa) and Miriam A. (Ma) Ferguson, to create the unique agency which parlayed a $5,000 loan into more than a billion dollars of assets in less than fifty years.

During this remarkable half century, five men served as general managers of the LCRA. Each served at the pleasure of the policy-making board of directors, whose members have been appointed by the governors of Texas since 1941 (prior to that time, the governor shared this appointive power with the state attorney general and the commissioner of the General Land Office). In theory, of course, the governors reflect the sentiments of the voters and the appointees the philosophies of the governors — thus providing citizen control over a board much more powerful than most people, including those in the area directly affected, probably realize.

The LCRA is a non-profit and, theoretically, a non-political agency with a unique status. It is subject to state control but receives no state funds. It is required by law to be self-supporting but has no taxing authority. Its only source of income is revenue from the water and electricity it sells, although one of the greatest benefits it provides the public undoubtedly is the prevention of floods.

Its by-products include a booming tourist industry along the 600-mile chain of Highland Lakes created by its six dams. However, the levels of Lakes Travis and Buchanan rise and fall at almost unbelievable rates throughout most "normal-rainfall" years. Waterfront owners along their shores curse the rice growers downstream whose demands for irrigation water periodically result in extremely low lake levels.

Many of these property owners do not realize that the rice growers were there first and, under the law, have first call on what is considered the river's "normal flow" of water. But what constitutes normal flow? That is a question as old, and as controversial, as the Colorado itself.

How old is that? No one really knows. Archaeological diggings in modern times have borne out theories about some of the primitive people who live along the river and whose descendants, the Lipans, Karankawas, and Tonkawas, greeted the Europeans who first landed in the area during the seventeenth century.

Ironically, that first landing appeared to be an honest mistake. According to John Henry Brown's *History of Texas, 1685–1892*, Réne Robert Cavelier, Sieur de la Salle, left France on July 24, 1684, with instructions to settle a colony at the mouth of the Mississippi River. Handicapped by a lack of adequate charts on the Gulf of Mexico, he sailed too far southward and entered what is now known as Matagorda Bay. There, near the present mouth of the Colorado, he established in February 1685 a small garrison and mission which he named Fort St. Louis.

Four years later, in April of 1689, an expedition led by Don Alonso de León, the governor of Coahuila, Mexico, reached what had been Fort St. Louis. The noted historian, T. R. Fehrenbach, wrote in his authoritative *Lone Star, A History of Texas and Texans*, that the expedition was greeted by rotting skeletons and desolation. What little was left of the fort at that time was burned to the ground, wrote Fehrenbach, and Don Alonso later learned that it had been devastated by a smallpox epidemic, then virtually destroyed by the Karankawas before he arrived.

Don Alonso decided to finish the job begun by the Karankawas. He was so successful that even the exact location of Fort St. Louis remained something of a mystery until the ruins were unearthed during the 1960s by archaeologists, according to Fehrenbach.

It was not until 1690 that the Colorado River first showed up on a map, and some historians believe that even that was a classic case of mistaken identity. Manuel José de Cárdenas y Magaña was the mapmaker on an expedition led by Captain Francisco de Llanos that year. The expedition reportedly moved up the river about ten or fifteen miles before being turned back by a mass of driftwood made up of trees and debris — a "raft" which would become famous in Colorado lore.

There is evidence to suggest that Cárdenas and another early mapmaker, Álvarez de Piñeda, used the name "Los Brazos del Mar" (The Arms of the Sea) or "Los Brazos del Dio" (The Arms of God) for the river now known as the Colorado, which is a Spanish word meaning "reddish." They may have intended that name for the siltier, red-tinted stream to the east, which now bears the name Brazos. But the theory is that early cartographers based in Madrid simply transposed the names by mistake, according to Brown's *History of Texas*.

Regardless of its name, what is now known as the Colorado River became a lifeline between 1680 and 1800 for Spanish explorers and missionaries, soldiers and settlers. They established missions and forts along the river, helping to set the stage for the Anglo invasion that was to be led by Stephen F. Austin.

His father, Moses Austin, was an adventurous lead mine operator from Virginia who moved to St. Louis, Missouri, in 1797, while that area was under Spanish rule. A skilled promoter, he became successful as a mine operator and gained the right to bring in thirty families from the United States. Later, after that area became a part of the United States, he established a bank which became highly prosperous; but Moses Austin was wiped out financially, along with his St. Louis bank, during the depression of 1818.

He then turned his attention to Texas and made a long, arduous journey to San Antonio. He cajoled the Spanish authorities there for permission to settle 300 families in Texas and finally wrangled a permit to do so. It was granted in January 1821, a few days after he left for St. Louis. But Moses Austin never saw the permit; he died before it caught up with him. He made a deathbed wish for his son to take over the project.

Stephen F. Austin, then only twenty-seven years old, immediately assumed his father's role as a colonizer and explored a wide area of Texas before picking the land for his initial settlers, who were to become known as the "Old Three Hundred." He went from Nacogdoches in the northeast to La Bahía (Goliad), southeast of San Antonio. The region which impressed him most lay between the Colorado and Brazos rivers, stretching all the way from the Gulf Coast northward to El Camino Real (an old military road connecting Nacogdoches with San Antonio). He succeeded in getting a grant for the rich river-bottom lands along the Colorado, Brazos, and San Bernard rivers.

According to Fehrenbach in *Lone Star,* Austin described this country as the best in the world, "as good in every respect as man could wish for, land first rate, plenty of timber, fine water — beautifully rolling."

Late in 1821, Austin brought his first immigrants into the areas around present-day Columbus, on the Colorado, and Washington-on-the-Brazos. He quickly encountered legal problems, caused by Mexico's winning its independence from Spain that same year, and it was not until 1824 that he established his first official colony headquarters at San Felipe de Austin.

The early settlers, mostly from Louisiana, Arkansas, Alabama, Tennessee, and Missouri, planted such crops as cotton, corn, feed grains, and vegetables in the fertile river bottoms. They cleared timber, built homes, hunted the native wildlife for food and furs, raised cattle and horses. Regardless of what they did, the Colorado River generally was the focal point of their existence.

They discovered, the hard way, that the river which provided life-giving water much of the time could go almost dry and then change suddenly into a raging torrent. With little or no warning, it could wash away most of their assets and even claim the lives of those who did not pay proper respect to it.

These settlers were among the first of many to suffer from the effects of the notorious raft — the mass of driftwood which choked the river channel in its lower reaches, blocking navigation and serving as a dam which spread floodwaters throughout much of the country they were trying to settle and cultivate. Ironically, they complicated the problem by permitting most of the brush and much of the timber they cleared from their land to wash downstream and reinforce the natural blockade.

With the raft making navigation of the river impossible, the settlers faced tremendous transportation problems. Mud resulting from normal rainfall and the frequent flooding made most of the primitive roads impassable for many months during nearly every year.

But as the raft continued to grow, so did the settlements along the stream. More and more people began to dream of shipping their produce to Gulf ports and receiving supplies on boats sailing up and down the Colorado.

In 1837, a year after Texas won its independence from Mexico and became an independent republic, these people took their case to the new country's Congress. They succeeded in getting a special law enacted on December 14, 1837, incorporating the Colorado River Navigation Company with the authority to issue capital stock amounting to $125,000 in shares of $100 each. The money would be used to clear the Colorado River channel for navigation and was to be repaid by tolls on steamboats and other craft using the river. The company was given four years to open the fifty miles of the channel choked by the raft.

This first attempt at government intervention in harnessing the Colorado failed miserably. Little, if any, money was raised, and no progress at all was made in eliminating the raft.

On June 1, 1842, delegates from the counties in the lower part of the Colorado watershed met in Columbus and made plans to raise $30,000 for removing the raft. The work was to begin when the first $10,000 was raised; however, two years went by without that magic mark being reached.

On January 18, 1844, the Congress of the Republic of Texas rechartered the Colorado River Navigation Company. That same year, Samuel Ward, a La Grange merchant, built a steamboat for the firm. He named it the *Kate Ward*. A side-wheel steamer, it was 115 feet long with a 24-foot beam and had two engines of 70 horsepower each. It had a shallow draft and could carry 800 bales of cotton, according to the late Walter E. Long's history of the Colorado's early development, *Flood to Faucet*.

The *Kate Ward* first arrived in Austin on March 8, 1846, just three months after Texas officially became the twenty-eighth of the United States. The City of Austin staged a great celebration in its honor, but the sturdy boat, built at the head of the raft, could transport goods only back that far down the river — roughly to the middle of Matagorda County.

Numerous other attempts were made, unsuccessfully, to raise money from private sources for removing the raft and making the Colorado navigable. These led, inevitably, to appeals for aid from the newly joined federal government. A U.S. Corps of Engineers report in 1851 noted that the raft was seven miles long, anchored by a solid bed of logs three miles long and varying from ten to twenty feet in depth.

That report, made by Lt. William H. C. Whiting of the Corps, has been credited with helping gain a $20,000 appropriation in the Rivers and Harbors Act of August 30, 1852, for "improvement of navigation of the Colorado River, Texas."

This was the first appropriation of money made by the Congress for the development of the Colorado. While it was barely enough to scratch the surface of the problem, it spurred subsequent interest by the Corps of Engineers. The federal government began, in November of 1853, the tedious and expensive process of clearing the river.

Instead of trying to remove the raft, the Corps of Engineers dug a by-pass channel around it and through a series of lakes. By March of 1854, a channel had been completed so that a boat could pass through. At long last, steamships could go directly from Indianola to Columbus and La Grange through a relatively safe channel.

Unfortunately, no funds were appropriated to deepen the channel, remove the snags upstream, or provide maintenance for the navigable channel. As a result, the heyday of Colorado River navigation lasted only about six years, from 1854 until 1860. Three steamships — the *Kate Ward,* the *Colorado,* and the *Betty Powell* — made regular trips between La Grange and Galveston during that period. Two smaller boats, the *Lareno* and the *Moccasin Belle,* made trips as far up the river as Bastrop.

Navigation of the Colorado was just becoming well established when it was brought to a screeching halt by the outbreak of the Civil War, the coming of the railroads, and the formation once again of the infamous raft.

Driftwood began collecting at the mouth of the new channel in 1859 and, without any efforts to remove it, navigation was blocked again in 1860. A devastating flood in 1869 produced substantial damages all along the river from Austin to Matagorda, contributing more debris to the raft. This, in effect, reinforced once again the natural dam which spread floodwaters over much of the lower Colorado's watershed. Several successive reports by U.S. Corps of Engineers officers declared that the river was not worthy of improvement for navigation purposes by the federal government.

But the flood of 1869 was just the forerunner in a series of disasters which eventually washed away any doubts that someone, somehow, had to bring under control the periodic rampages of the mighty Colorado — navigation or no navigation.

That process took more than seventy-five years. Even though Austin, the state capital, bore the brunt of many destructive floods during that time, it finally took the skillful political log-rolling of the Governors Ferguson to sire the Lower Colorado River Authority. The Fergusons managed to trade Texas's 100th birthday party "cake" for the mechanism needed to control the wild, often-furious Colorado — paving the way, ultimately, for Austin's phenomenal growth.

2

Dirty Tricks

Almost as old as the Colorado's flooding problem is that of securing enough water to irrigate crops in the Lower Valley. Compounded by Mother Nature's erratic whims, the region produced a crazy-quilt pattern of human behavior for many years. While people banded together and usually tried to help each other during floods, they fought fiercely at other times over what amounted to a mere trickle of water. The high hopes of many early settlers were dashed to bits by the lack of adequate water supplies for irrigation.

The first residents started trying to grow sugar cane as early as 1829. In 1831, Stephen F. Austin even attempted to lure additional settlers to his colony by advertising that it was located in the midst of a 70,000-square-mile area well suited for growing sugar cane.

Railroad promoters echoed his glowing predictions about the future of sugar production. After the first sugar mill was built in Brazoria County during the early 1830s, they dubbed the region covered by Brazoria, Matagorda, and Colorado counties the "Sugar Bowl," according to S. G. Reed's respected book, *A History of Texas Railroads.*

When the floods came, of course, it mattered not whether the crops were sugar cane, cotton, feed grains, kitchen garden vegeta-

10

bles, or even rice, which eventually would become the area's top money-producer. All were washed away equally by torrents of water which, ironically, sometimes rekindled the ill-fated hopes for navigation of the Colorado.

Historians seem to agree that the worst flood in the history of the Colorado was that of July 1869, when the water rose to a height of forty-three feet at Austin. The flood stage was twenty-one feet. At its peak, the flood sent an estimated 520,000 cubic feet of water per second cascading down the Colorado.

Dr. Comer Clay, in his excellent 1948 dissertation on the LCRA, noted that "The immense power of such a flood can hardly be imagined by the average person." He wrote:

> The flow of Niagra Falls varies from 153,000 to 267,000 cubic feet per second from standard low water to standard high water. By comparison, it is seen that the 1913 flood at Columbus equalled and the 1869 flood at Austin was twice as great as the high flow of Niagra. The discharge rate of the 1869 flood at Austin was as great as the average flow of the mighty Mississippi River at Vicksburg.

Frank Brown, who moved to Austin in 1846 and served as county clerk and later as district clerk of Travis County, wrote an unpublished manuscript entitled "Annals of Travis County and the City of Austin," which is now in the Texas State Archives. He quoted an unnamed eyewitness of the 1869 flood as saying:

> The river reached its highest mark on the evening of July 7, at about 9 o'clock. The rise was estimated at forty-six feet. The mass of waters rushed down from the narrow and confined channel between the mountains above, to the wider one below, with such fearful velocity that the middle of the stream was higher than the sides and the aspect it presented was appalling.

According to the July 14, 1869, edition of the *Tri-Weekly State Gazette,* published in Austin, that flood put Bastrop under water and covered the business district of La Grange with six to ten feet of water. It inflicted severe damage on crops and buildings throughout the river's basin; but still, the determined farmers who suffered most apparently never gave any serious thought to throwing in the towel. Instead, they simply went back to work, salvaging what they could and preparing to plant new crops.

It was not until the 1890s that the farmers began to realize what vast profits might be reaped from raising rice if they could exercise even a moderate amount of control over the unruly waters of the Colorado.

Capt. William Dunovant, a native of Chester, South Carolina, came to Texas in about 1874 and began raising sugar cane near Eagle Lake. When he heard that an irrigation system using steam-driven pumps had been devised for rice production near Crowley, Louisiana, in 1885, he decided to plant forty acres of rice. He devised a crude irrigation system, which turned out to be little more than a ditch connected to Matthew Bayou. But apparently it was successful enough to make him expand it and also prompt some of his neighbors to build similar canals.

These pioneering efforts resulted in the creation of many irrigation companies along the lower Colorado, enabling the early-day firms to claim permanent water rights which would complicate development of the river for many years. In the early stages, their "water rights" also complicated their relations with each other. And for many years, they all could laugh and joke and be thankful that their "aim was bad."

Earl Eidelbach of Bay City, who retired from the LCRA's Gulf Coast Water Division in January of 1986, grew up around, along, and with the various irrigation canal companies which made Texas one of the nation's top rice-producing states. His father went to work for one of the irrigation companies in 1909, but Earl waited until 1929 — when he was nine years old — to start his career. He began as an errand boy for a canal company and worked his way up to superintendent of the LCRA's Gulf Coast canal operations.

He knows the canals and the canal companies inside and out, from personal experience and from family history. No one appreciates more than he does the importance of water, especially to rice farmers. In a 1985 interview he recalled:

> Back in the early days, rice farming was very crude and mostly was done with mules, horses, and manual labor. Water was very insufficient at times. We depended on rain and the help of the river. There was, at one time, around twenty-five canal companies in Matagorda County. They later consolidated into one irrigation system.
>
> The times were hard back then. They tried to build the canals large enough to make small reservoirs out of them. They built some of them one hundred and fifty feet wide, so when the

river was on a rise they could pump and store enough water to get them through a short dry period. But their timing had to be just right for them to make a good crop.

There were times when we had a drouth and the rice was hurt real bad. Then we'd have a flood and it would wash all the levees away . . .

This was back in the days when everything was done the hard way, and the rice crop was their life existence. It was their meal ticket and they had to have it to exist, so they went to any means and manner of doing the best they could to provide for their families.

The families would often move from nearby towns into camps near the rice fields to provide the intensive labor needed at planting time. Water also was needed at precise stages in the growing cycle, and in fairly predictable amounts, but it was not always as easy to obtain as was labor. Eidelbach recalled:

Sometimes, when the river would go dry and they didn't have any water, everybody was fighting over it. The man on the upper end of the canal was going to get his when it came by and the man on the lower end was threatening to shoot him for taking more than he ought to and not letting it come on down.

Actually, there was a lot of gun-play but I don't know that anyone was ever shot. I've heard of 'em being shot *at* — but not being hit. Everybody carried a gun and hoped that nobody would use one. But they did use dynamite.

Explosives were a tool of the trade, he noted, at a time when there were about twenty-five privately owned canal companies operating along the lower Colorado. All of those firms were fiercely competitive and willing to do just about anything necessary to get water for the rice crops.

These companies tried to cut one another's throats to make them stop their pumps, so the next man down the river could get water when there wasn't very much water in the river. They'd go out and dynamite their flumes and other structures, such as under-drains. They'd have to shut down to make repairs and while they were shut down, this other canal system would get their acreage watered up. There was always a fight for water back in the old days because water was the main thing you had to have to raise rice.

Eidelbach started his career as an errand boy for Victor L.

LeTulle, who had built the Bay City Irrigation Company in 1901. LeTulle's early success sparked the growth of competition and, by 1916, irrigation systems were spreading water over 145,000 acres of land in Colorado, Wharton, and Matagorda counties. Twenty-seven pumping plants, of varying sizes and ownership, were operating in the three counties; they had a total pumping capacity of 2,512,000 gallons per minute, but rarely had enough water available to utilize it fully.

All of the plants exercised some claim to the waters of the Colorado, although many did not last long. In 1927, LeTulle bought the other twelve canal companies then operating in Matagorda County and one in Wharton County, combining them into one of the largest individually owned irrigation systems in the world. He named it the Gulf Coast Irrigation System.

In March of 1931, LeTulle sold his firm to a utility company, Interstate Public Service, which renamed it the Gulf Coast Water Company. By 1934, when a severe drouth all but dried up the river, Gulf Coast was one of three irrigation companies still operating. The three made a deal for water with the Brown County Water Control and Improvement District, in West Texas, agreeing to buy $10,000 worth of water from Lake Brownwood. A system of daily telephone communications was established so the water would be released only in amounts capable of being pumped in the three counties involved. In addition, an elaborate policing system was arranged to make certain that others did not take the water and that none of the paying participants received more than their authorized allotments.

About the time the last of the contracted water was released, officials of the Brownwood District began trying to get the irrigation companies to buy more. Although they needed more water, the companies kept insisting they could not afford it and reluctantly turned down the offers. But, said Eidelbach,

> the water kept on coming for another week or ten days, and we got quite a bit more water than we had bargained for. It was a real good deal for us. But it turned out that the gate got stuck open at the Brownwood Dam. They couldn't get it closed and they were afraid it was going to drain their lake dry.
>
> Meanwhile, the Gulf Coast Water Company had built a dam of sand bags across the river below Bay City to try and hold all of this water they could. But with this gate getting stuck up there,

there was just more water coming down the river than they could pump, and it washed their dam away — so they had a losing factor, too.

[The irrigation canals were laced with locks and "water boxes" to control the flow of water into various farms and fields.]

Farmers used to always be going down there at night to block somebody else's lock and open the water boxes into their own fields, to steal all the water they could. The next morning, the water boss [a supervisor for the canal company] would catch it and have to go set it back like it was. But there was no way you could prove who did it. You *knew* who did it, of course, but they'd always deny it and unless you saw 'em do it, you couldn't prove it. They'd say they just didn't know who could've done that.

The farmers were all good friends and when they got through, in the fall of the year, and everybody had harvested their crops, they'd sit around and hooraw one another. They'd shoot the bull and tell what they'd done to one another to steal their water. They were good friends but when it came down to water, everybody was for himself. That was just a way of life.

But one time, at least, they all worked together in a clandestine scheme to get the water they needed, according to Eidelbach, and they even defied the Texas Rangers.

By 1920, the raft of logs, trees, and brush had built up between Bay City and Wharton to such an extent that it backed up floodwaters throughout much of the two towns. A 1923 act of the Texas legislature rescinded state ad valorem taxes for twenty-five years in Wharton County and three precincts of Matagorda County to finance bonds, thus raising money to clear out the raft and to build levees and drainage canals.

According to Dr. Comer Clay, Matagorda County voted bonds amounting to $685,000 and Wharton County approved bonds totaling $540,000 for this purpose. The work started in 1925 and by 1928 a channel had been opened through the entire raft. But the river almost went dry, Eidelbach remembered,

and the rice crops were burning up. There was a dry creek up in Wharton County, on the west side of the river, and the dredging company that cleaned out the raft had left an old dredge up there. LeTulle got the farmers and a bunch of other people together and commandeered that old dredging barge. They cut holes in it with axes and sunk it, then dumped trees and brush behind it to build a makeshift dam.

Well, when the Wharton County people found out about it, they got the Texas Rangers to come out there and arrest everybody — including my Dad — and take 'em to the Wharton County Jail. LeTulle had a lawyer, Will Davant, put up bond for everybody and got 'em released. And then the sheriff of Matagorda County, Joe Mangum, told them all to go on back and finish their dam.

They went back to work and every day the Rangers would come out and arrest the crew — but there was always another crew hiding out in the river bottom, and it would take over and go right on with the work. They managed to get enough water to save their rice crop, and nobody was ever prosecuted. As a matter of fact, the Rangers said later if they had known it was just rice farmers trying to get water for their crops, they'd have never come down.

A severe flood during the summer of 1929 finally conquered the raft, knocking it out of the Colorado River but into Matagorda Bay, where it created both trouble and land.

Norman Treude, manager of the LCRA's Gulf Coast Water Division, noted that each succeeding rise brought more logs and debris down the river, building up the raft until it actually joined the mainland with Matagorda Peninsula.

"They finally had to go in there in 1934 and dredge an opening, about two hundred feet wide, through all that debris so the water could get out into the Gulf," said Treude, pointing out that every rise coming down the Colorado extends the man-made beach of Matagorda Bay a little more.

Treude, a Texas A&M graduate who joined the LCRA in 1960, supervises the irrigation of about 40,000 acres of rice — in an era where scientific methods have replaced the need for the trickery and deceit which farmers once used. Farmers now use laser beams to run contours in planting so that they can gain maximum benefits from the water they get legally.

"But we have to order our water a week in advance, since it takes that long for it to come down the river," said Treude. "We have to know today what we are going to be pumping a week from now — whether it rains or not. We just assume it's *not* going to rain."

Farmers usually get two rice crops a year.

"For years, we have figured that it takes seven acre-feet of water per acre for a farmer to grow two rice crops a year," he said, "and our water charges are based on that figure." (One acre-foot is the amount of water needed to cover an acre of ground to a depth of one foot).

To a rice farmer, that seems a reasonable amount of water for a crop that means up to $300 million annually to the economy of Texas. To the vacationers who flock to the chain of Highland Lakes — especially Lakes Travis and Buchanan, where the drastic fluctuations in levels are caused so often by the rice farmers' needs — it sometimes seems outrageous.

The low levels which occur periodically provoke more criticism from occasional visitors to the Highland Lakes than they do from those who live near them or vacation there regularly. During dry summer months, when Lakes Travis and Buchanan frequently are low, many visitors are likely to complain that "the rice farmers got all our water."

But the rice farmers' claims on Colorado River water date back to the turn of the century, and many of them were in the forefront of the long crusade to cage the Colorado. It took a lot of political clout from many different sources to create the Lower Colorado River Authority. Much of it came from the rice farmers.

It is true, as Treude says, that the building of the LCRA dams "made rice farming a sure thing" instead of a high-stakes gamble every year on the weather.

Treude and his family have been vacationing regularly for thirty years at Kingsland, a resort community on Lake LBJ, which happens to be one of the reservoirs maintained at a constant level throughout the year.

"But when I'm up there," said Treude, grinning, "I still don't volunteer the information that I'm from down here in the rice country."

3

A Power Struggle

Traces of the Colorado's longstanding courtship of controversy can be found even in the selection of Waterloo, later renamed Austin, as the permanent capital for the Republic of Texas. Five commissioners appointed under an 1839 act of the Republic's Congress apparently agreed that the capital should be located either on the Brazos River or the Colorado. By a vote of three to two, they named Waterloo in their report to President Mirabeau B. Lamar, citing "the greatest and most convenient water power to be found in the Republic" as one of its main assets. Congress ratified the commission's action on January 19, 1840, starting a "power struggle" that still continues.

By 1840, two water mills were operating in what is now downtown Austin, according to Frank Brown's *Annals of Travis County*. Both were wiped out by sudden floods. Jonas Dancer, a Methodist minister, built another mill four miles above Austin; between 1850 and 1859, it was washed away three times and rebuilt three times by the indomitable Dancer. In 1855 a floating grist mill was built in Austin just below the mouth of Shoal Creek and was anchored by a cable attached to a large tree. After it also was swept away by

18

a flood, Brown concluded: "Experience has demonstrated that the Colorado is too treacherous for mills."

For a long time, especially to the citizens of Austin, the river also seemed too treacherous for dams. Professor Thomas U. Taylor, who was dean of engineering at The University of Texas, wrote a 1904 report on *The Water Powers of Texas.* In it, he cited one successful, though rather modest, pioneering effort to build a dam across the Colorado in 1879 near Lometa. Built by Milam Chadwick, it was made of stone and cedar logs, measured 196 feet long, and gave the water a clear fall of five feet. A mill race enabled the falling water to turn a thirty-six-inch turbine. Chadwick cleverly used a wire cable to transmit the power to his mill, which was high on the bank and thus out of danger from high water. He milled flour and corn, and also had a cotton gin; Chadwick's Mill became a landmark for early settlers in Central Texas and was still operating as late as 1900.

But the first attempt to build a large dam and reservoir in Texas was launched by the City of Austin. In 1871, Mayor J. W. Glenn took the initial step when he directed the city engineer to make surveys of possible sites. Two years later the city granted a charter to a private company to build a dam across the Colorado, but the firm never started the project.

In 1888 the Austin Board of Trade hired J. F. Pope, a civil engineer, to make surveys for a dam. Pope made a report strongly supporting the construction of a dam and wrote some articles on the subject, one of which was dated October 14, 1889. The report appeared on October 31 in the same issue of the *Austin Statesman* in which a candidate for mayor, John McDonald, announced that he also favored building a dam.

The next week, the newspaper's editor commended McDonald for supporting the project and asked in print if the incumbent, Mayor Joseph Nalle, also favored it. When Nalle declared that he had grave doubts about the wisdom of building a dam, the *Statesman* endorsed McDonald. An editorial in the paper said the dam should be built due to

> the necessity of the City freeing itself from the grip of the existing water and light company, that is charging for these necessities five times what they can be produced for. This can best be accomplished by the City damming the Colorado.

On December 4 of that year, McDonald and a city council which favored construction of the dam were elected. Nalle became president of the privately owned Austin Water, Light and Power Company, and he continued to campaign against building a dam.

In February of 1890, Joseph P. Frizell, a civil engineer, was employed to plan the proposed project. In a report dated March 26, 1890, he recommended a dam sixteen feet wide at the top and fifty feet wide at the base, to be built of limestone and granite at an estimated cost of $1.3 million.

The City Council created a Board of Public Works to supervise construction and, on May 5, 1890, the voters approved the issuance of $1.4 million in bonds, paying five percent interest, for the daring but ill-fated venture. The vote was 1,354 to 50. Because of its size, 1,235 feet in length with a height of 60 feet above low water, the project attracted national attention — especially since Austin's population was only 15,000.

Determined that the job be done right, the board hired Frizell as chief engineer, Pope as first assistant engineer, and John Bogart, a civil engineer from New York, as a consultant. Bogart made a report concurring with Frizell's previous findings. His concluding statements were most reassuring to the citizens of Austin:

> The City will have, if it constructs this dam, a constant power equal to either of the above places [Lowell, Massachusetts, and Manchester, New Hampshire]. That there will follow a development of industries greater than one cares to anticipate seems to me assured, if this great water-power be assured here in the midst of this fertile region.

The people of Austin were eager to obtain not only their own stable source of water, at economical rates, but also power to light the city and operate its streetcars.

A contract for the construction of the dam was let to Bernard Corrigan of Kansas City, Missouri, on October 15, 1890, and excavation was started on November 5.

Again exercising an apparent abundance of caution, the Board of Public Works engaged J. T. Fanning, a Minneapolis engineer, to inspect the dam and the proposed site for the power house. In his report of June 1892, he recommended a few changes and compared the project with other large dams in California, France, Belgium, and Spain. He noted that some of those exceeded the Austin Dam in height; however, none of these dams are upon great rivers, and very few of them have any water pass over their crest. On the other hand, the Austin Dam stands in the channel of the Colorado River, where it has 40,000 square miles of watershed, and will have floods of 200,000 to 250,000 cubic feet of water per second to pass from its crest to its toe. Your citizens will appreciate your responsibility when they learn that no dam in existence has to pass a volume of water, in flood, even approximating this, through so great a height. Limestone and sandstone yield rapidly to the eroding force of falling water.

Controversy dogged construction of the dam. It was being built on a geological fault, and foundation difficulties developed quickly. Those were followed by a rapid turnover of engineers, many of whom resigned with complaints that they could not get their instructions followed and that construction was not meeting the specifications.

Meanwhile, the sale of the bond issue failed to meet expectations. In a report to the City Council on October 2, 1892, Mayor McDonald reviewed the history of the project and denounced the Austin Water, Light and Power Company, which was still supplying the city with water. Recalling the passage of the $1.4 million bond issue, he declared:

> The friends of the existing water company at once began a warfare upon the City which for malice, meanness and mendacity has rarely been equalled. They used every means in their power to retard our work and prevent, if possible, the sale of our bonds. They flooded the country with anonymous communications assailing the credit of our City, and attacking the integrity of our citizens . . . They hired known villifiers to write false statements and scandalous articles concerning us . . . But for this bitter and malignant warfare on the part of our enemies, we would have sold the remainder of our bonds a year ago . . .

The last $388,000 of the bond issue finally was sold in 1893 at ninety-two cents on the dollar.

The dam was completed on May 2, 1893, and reportedly was the largest in the world across a major river at that time. Two weeks later, Lake McDonald — named in honor of the mayor — had filled and water began flowing over the crest of the dam.

But before the end of the month, the first in a series of misfortunes struck. On May 30, 1893, water from the lake seeped through a soft stratum of rock under the dam and into the excavation for the power house. That defect was corrected rather quickly but proved to be only the forerunner of other problems which plagued this pioneering effort to harness the Colorado.

The power house was substantially completed by March 7, 1895, when the pumps of the city waterworks were started. A backlog of applications for city water and power connections built up quickly. On May 21, the City Council called an election for June 24 to issue another $200,000 in bonds for extension of the distribution systems and to pay off $50,000 indebtedness on the power plant. Mayor McDonald declared:

> Our waterworks, electric light and power plants have been completed and are in successful operation. The City of Austin is the owner and operates the best water power, waterworks, electric light and electric power plant owned by any city in the United States, and we have a right to feel proud of our achievement.

According to the *Austin Statesman* of June 25, 1895, the new bond issue was approved by a vote of 1,219 to 115.

That seemed to solve the financial and political problems, but the engineering problems surfaced again and again. The engineers had seriously overestimated the minimum flow of the Colorado, underestimated the amount of water that would be lost to evaporation, and failed to anticipate that 31,667,000 cubic yards of silt would flow into the reservoir during its first four years, taking up 38 percent of its original capacity.

At first, Austinites enjoyed city water and electricity at rates that were about half of what the private company charged. City streets were lighted by arc lamps atop 31 towers, each 150 feet high, which produced a brilliant moonlight effect and made Austin known as "The City With the Violet Crown." Electricity provided power for streetcars, which previously had been drawn by mules.

Lake McDonald quickly became a popular recreational center. Shows presented at a lakeside pavilion and grandstand drew large crowds during the hot summer months. And the double-decked *Chautauqua* could carry as many as 500 people on moonlight cruises and sightseeing trips, according to David C. Humphrey's book, *Austin — An Illustrated History*. Humphrey wrote:

> But the queen of the fleet was the *Ben Hur*, a side-wheeler with a capacity of 2,000 passengers. It had three decks, twenty staterooms,

fruit and lunch stands, a nursery, a laundry, electric lights, and a touring company that performed vaudeville and melodrama on its decks. University students, firemen, Masons, conventioneers, and dozens of other private parties booked excursions on the lake. Thousands of individuals took advantage of the regular daily schedule of sailings. The crowds were often very lively, and it was not unknown for a voyager who had piloted too many schooners of beer into port to take a header into the water . . .

The party ended almost as suddenly as it had begun, when the stark realities of dry weather caught up with the overly optimistic estimates which had been made originally. By 1899, it became obvious that the dam could not be depended upon to supply essential water and electricity to Austin.

Humphrey's account continued:

During January [1899] the lake level dropped steadily. A power shortfall began to develop. On January 22 current for the tower lights was cut off — not [to] be turned on again for three months. By March 9 the lake was almost nine and one-half feet below the crest of the dam. Service to the city's private electric customers was stopped, except for those on Congress Avenue. Some turned to the private utility company, whose steam plant continued in operation. The lake level continued to fall. On March 20 the streetcar system was shut down entirely — for a month, it would turn out. The Austin Rapid Transit Company hired thirty-eight mules, restoring the slow mule-drawn service that electricity had supplanted several years earlier . . . On March 29 the city ceased providing electricity altogether and announced that it did not expect to resume service until there was a heavy rainfall. Customers were advised to rely on candles and oil lamps. Not until mid-April did the city dam generate electricity again.

In this instance the city furnished water despite the power shortage. In September residents were not so lucky. Once again the lake level dropped, the city stopped generating electricity, and the mules were trotted out — this time for six weeks. Water service from the "city dew plant" was also halted. When the old temporary capitol building burned on September 30, firemen had to use the private company's hydrants. Fire insurance companies threatened to raise rates unless the city came up with a reliable water supply.

By February of 1900, most of the water had disappeared and,

according to Professor Taylor, the reservoir was forty-eight percent filled with silt.

And then came another of those torrential rainfalls throughout the Colorado watershed, producing the greatest flood on the Colorado since 1869. Five inches of rain fell in the Austin area during the fifteen hours beginning at 1:00 P.M. on April 6, 1900. Heavy rains also pounded the Colorado area and its tributaries for about 100 miles above Austin. By 10:00 A.M. on Saturday, April 7, water was flowing about ten feet above the crest of the Austin Dam. Ten months earlier, on June 7, 1899, the massive structure had weathered a flood that sent water cascading almost ten feet above the crest. But this time, the water won. Professor Taylor was there when the dam broke and later reported:

> At 11:20 a.m. on April 7, when the lake level had reached a height of 11.07 feet above the crest of the dam, the dam gave way . . . about 300 feet from the east end of the dam . . . Sooner than it takes to write these words, the two sections — each about 250 feet long — were shoved or pushed about 60 feet from their former positions. There was not the slightest over-turning . . . By 11:30 a.m., the lake level had fallen to the crest of the dam. Measuring along the crest of the break, 456 feet of the dam to the west and 83 feet to the east was left still standing, unaffected.

The two portions of the dam which had broken loose remained upright for forty minutes before turning over and disappearing beneath the flood waters which rushed down the Colorado River to Bastrop, Fayette, Colorado, Wharton, and Matagorda counties.

Widely hailed as a hero of the disaster was O. D. Parker, manager of the Western Union Telegraph Company in Austin. He was credited with saving a great many lives by quickly wiring word of the dam's collapse to every telegraph operator between Austin and the Gulf of Mexico, according to Walter Long's *Flood to Faucet*.

Long served from 1915 until 1949 as manager of the Austin Chamber of Commerce and from 1915 until 1937 as secretary of the Colorado River Improvement Association. He apparently was among the first to realize the vast recreational potential that could result from taming the Colorado. When he wrote *Flood to Faucet* in 1956, Long cited the "entertainment and recreation features . . . and glamorous events which took place on and around Lake McDonald in the 90s, when entertainment was more limited than today." He continued:

The momentum for this river development, so badly needed later, stemmed from happenings of those days. Throughout the spring and summer months excursions by rail were run every weekend to Austin from as far away as one hundred miles in each direction by the three railroads serving Austin — The International and Great Northern, the Southern Pacific, and The Austin North-Western. Canoeing on the peaceful waters of Lake McDonald was of course a common sport, while numerous sail boats were kept along the shores near the dam. The *Ben Hur,* a two-funneled steamer of really masterful proportions, carried from 200 to 500 excursionists from the docks near the dam up the lake for many miles . . .

Although engineers disagreed on the cause of the dam's collapse, Professor Taylor wrote in 1904 that the consensus of engineering opinion was that it would have survived had the toe of the dam been "countersunk" into bedrock and if a concrete apron had been built below it to prevent "undercutting."

Immediately after the break, the Austin City Council, encouraged by an anxious citizenry, began plans for restoring the dam. Soon after the dam's failure, the Caswell Cotton Mill loaned its steam facilities to the city and several dynamos were brought to Austin for connection to its steam plant. This took care of a part of Austin until 1901, when the city bought from Nalle's company its city water system and constructed the first unit of a small electric generating plant.

People living downriver from Austin saw in the collapse of the dam another roadblock in their long efforts to produce dependable water supplies for irrigation and protection from devastating floods, as well as to seek out the possibility of navigation.

It took a long time to overcome that roadblock. Dr. Comer Clay noted that the City of Austin finally decided not to assume the financial and administrative burdens it had in 1890 but entered into a contract with William D. Johnson on September 22, 1911, to rebuild the dam. That contract called for Johnson to finance the construction program. The city agreed to pay $100,000 upon acceptance of the completed dam and power plant and also to make semiannual payments of $32,400 for twenty-five years. This would obligate the city to a total expenditure of only $1.72 million.

The actual construction work was subcontracted to the William P. Carmichael Company. The project was nearing completion when another disastrous flood struck on April 23, 1915, dealing it a near fatal blow. A short time later, the Austin City Council learned that some unauthorized changes had been made in the specifications and refused to accept the dam. The crowning blow came in September of that year, when still another flood roared down the Colorado and swept away the last hopes for successful completion of the project. Three months later, the Carmichael Company went into receivership.

The City of Austin finally hired Daniel W. Mead, a consulting engineer from Madison, Wisconsin, in 1917 to examine the contract it had made with Johnson. His subsequent report declared that the city had acted properly, under the circumstances, in refusing to accept the dam because of the alterations in specifications. But it was not until 1928 that the City of Austin filed suit to cancel the contract and recover its property.

In the meantime, Johnson's contract had been transferred to Austin Dam, Inc. The City of Austin made a new agreement with that firm to complete the dam; however, when it failed to take any action, the City Council voted on December 1, 1933, to declare that contract null and void.

Thus the City of Austin took title once again to the crushed remains of what had appeared to be the brightest and most promising hope ever devised for taming the Colorado. And it became evident that the age-old dream of conquering this fickle, ferocious river could be achieved only in a much broader political arena — one with huge financial resources.

4

The Pyramid Falls

Austin's spectacular but ill-fated attempts to dam the Colorado attracted enough publicity to focus renewed public attention on the incessant problem which men had been trying to solve for many years. A massive dam across the unruly river had been proposed at least as early as 1854, when a twenty-year-old surveyor named Adam Rankin Johnson marked a suggested site. Buchanan Dam eventually was built at that spot, and it stands as something of a monument to his foresight.

Johnson, born in Kentucky, was surveying school lands for the State of Texas when he crossed the Colorado at what was then known as Shirley Shoals. He was so impressed with the possibility of building a dam there that he chiseled a cross on a granite boulder to mark the location.

Later, Johnson fought in the Civil War and became a brigadier general in the Confederate Army. On March 23, 1865, in the Battle of Grubbs Cross Roads, Kentucky, he was blinded by a fateful shot that slashed across the bridge of his nose. The wound destroyed his sight — but not his vision. He returned to Texas and entered the real estate business in Burnet. He later founded the city of Marble Falls and built a textile mill there, relying on water power from the falls on the Colorado River for which the town was named. The mill failed, but General Johnson could still see the vast possibilities offered by the Colorado.

In 1885 he and a friend, Tom Chamberlain, the county surveyor of Burnet County, went back to the spot where he had marked that cross on the boulder. Johnson's memory, undoubtedly refreshed by Chamberlain's description of the scene, enabled him to dictate details of his proposed dam. Chamberlain sketched it for him, and General Johnson used the drawing to help gain what later proved to be valuable water rights at the Shirley Shoals site. Johnson bought ten-acre sites on each side of the river; today, Buchanan Dam splits those two tracts almost directly in half.

During the early 1900s, General Johnson managed to interest Charles H. Alexander, Sr., a wealthy Dallas businessman, in the possibility of building a series of power dams along the Colorado. In 1909, Alexander actually began building a dam, which was to be 785 feet long and 63 feet high, about one and a quarter miles below Marble Falls. He also made his own survey of the Colorado, from near San Saba to Austin, and he bought and sold water rights along the river. However, most of the fortune he owned at one time reportedly was washed away by his attempt to build the big dam. The project was kept afloat for a few years with the aid of his sons, Charles H. Alexander, Jr., and Jay Alexander, and his son-in-law, Walter McAllister of San Antonio.

About one-half of the dam had been completed when the project had to be abandoned due to lack of funds. Its remnants, representing an investment of several million dollars (in pre-World War I money) now lie beneath the waters of Lake Marble Falls.

Before General Johnson died in 1922, he sold his Shirley Shoals water rights to Alexander. Still determined, Alexander had assured the State Board of Water Engineers, in a letter dated June 11, 1919: "I can and will build a large power and irrigation project which will be of untold benefit to a large portion of Texas."

Alexander organized the Syndicate Power Company, with C. H. Alexander, Jr., as president and his other son, Jay Alexander, as secretary. In 1926 that firm sought permits from the State Board of Water Engineers for the construction of six dams on the Colorado — two in Travis County and four in Burnet County.

A bitter fight over water rights erupted in 1927, when the Brown County Water Improvement District No. 1 applied for a permit to build a dam near Brownwood that would impound 500,000 acre-feet of water. The Brown County application sparked what turned out to be a long-running feud between water interests in the upper and lower parts of the Colorado River Valley.

In the meantime, John B. Carrington of San Antonio had acquired water rights along much of the Colorado below Austin. The Austin Chamber of Commerce urged him to help the Alexanders' Syndicate Power Company get the Insull brothers of Chicago interested in the possibility of building a series of dams above Austin. Martin J. and Samuel Insull owned the Middle West Utilities Company, which controlled several electric utility firms including the West Texas Utilities Company, headquartered at Abilene, and the Central Power and Light Company, which had its home office in Corpus Christi.

"Insull" was a highly respected name in the field of power development during the early years of the twentieth century. Samuel Insull, who founded the utility empire, was an English bookkeeper who had caught the eye of Thomas Alva Edison while the famous inventor was visiting in London. Insull came to the United States in 1881 as Edison's secretary and quickly began a remarkable climb up the corporate ladder. He eventually became president of the Edison General Electric Company and later headed Chicago's Edison Electric Company, a pioneer electric utility. He became a U.S. citizen in 1896. His innovative financing techniques, which utilized holding companies and a wide array of stock promotions, stirred up a great deal of controversy but did nothing to diminish respect for his ability and his financial muscle. But that was before the Great Depression, when he finally went bankrupt and fled the country with angry stockholders yapping at his heels.

In 1927, however, he saw great possibilities in the power of the treacherous Colorado River. Working through their Central Power and Light Company, the Insulls made preliminary plans to build at least five dams across the Colorado, and perhaps a sixth, at a cost of approximately $15 million. But at a mass meeting in Abilene on August 18, 1927, called by the West Texas Chamber of Commerce, the Insulls were charged with trying to "rob the people of West Texas of their water supply." A resolution was adopted urging the Board of Water Engineers to grant the Brownwood permit; eventually, that permit was granted, leading to the completion of the Brownwood Dam in 1932.

Many of the West Texans waged an acrimonious battle against the Insulls, fearing that construction of the proposed dams might eventually claim water they would need to support increased population. The Insulls, declaring that the goodwill of people in that area was essential for the West Texas Utilities Company, decided to drop their plans to enter the dam business. "I didn't come to Texas to get into a lawsuit," said Martin Insull. But, almost immediately, they were besieged by people downstream from Brownwood urging them to reconsider. Those people felt the Insulls offered their best hope for providing protection from floods, for establishing dependable water supplies, and for constructing badly needed power plants.

Two Texas congressmen who were destined to play key roles in the ultimate taming of the river, Joseph J. Mansfield of Columbus and James Paul Buchanan of Brenham, were among those trying to persuade the Insulls to build six dams above Austin plus another near Columbus. Both men had great influence: Buchanan as chairman of the Appropriations Committee in the U.S. House of Representatives; Mansfield as chairman of the Rivers and Harbors Committee. The Colorado River Improvement Association, composed of prominent and influential citizens throughout the lower valley, adopted a resolution asking the Insulls to reconsider. A. J. Eilers, its president, declared that the Insulls and the Syndicate Power Company should build the dams to prevent floods which he said were costing an average of $4 million a year, to irrigate another 400,000 acres of land in the lower valley and to furnish "an immense amount of electric power without the consumption of one drop of water."

The West Texas Chamber of Commerce continued, however, to spearhead determined opposition to the downstream dams and won editorial support from newspapers in San Angelo, Abilene, Fort Worth, and Dallas. But the Chamber's Special Water Rights Committee met in Fort Worth on February 13 and 14, 1928, with representatives of the Syndicate Power Company and worked out an agreement that seemed to satisfy everyone — except the Insulls. Syndicate Power owned permits for 1,225,000 acre-feet of Colorado River water. Its representatives agreed to give up 645,000 acre-feet, provided their firm was purchased by the Insulls. West Texas cities which had protested so vehemently felt that would assure them of adequate water supplies, and the Chamber's Board of Directors approved the agreement.

Details of the agreement were presented to the organization's annual convention in Fort Worth on June 18, 1928. By then, the Insulls had spent an estimated $500,000 in purchasing water rights and making surveys, according to the *Austin American* of April 19, 1931. But Martin Insull, president of the Middle West Utilities Company, apparently was tired of being caught in the crossfire between the upstream and downstream interests and perhaps felt that their truce would not last. At any rate, he notified the West Texas Chamber of Commerce on July 24, 1928, that his firm had dropped all plans to build dams on the Colorado.

That came as a demoralizing blow to the people in the lower valley, all but demolishing their hopes for using the river instead of suffering from it. But hope — indeed, a savior — appeared a short time later in the form of State Senator Alvin J. Wirtz of Seguin. He was slated to become known as a brilliant, crafty lawyer, as one of Lyndon B. Johnson's closest friends, and as a representative for the eventually powerful Brown brothers, George and Herman.

Wirtz represented a wide variety of clients, including two of the biggest oil companies in Texas — the Humble Oil and Refining Company and the Magnolia Petroleum Company. He also represented the Garwood Irrigation Company of Colorado County, one of the many small canal companies which supplied water to rice farmers whenever it could be obtained from the erratic Colorado. More importantly, one of his major clients was a Chicago financial firm, Emery, Peck and Rockwood, for which he was playing a key role in the construction of six small dams on the Guadalupe River. Wirtz was primarily responsible for acquiring the water permits and reservoir lands for those dams, which were designed to generate hydroelectric power and provide water for irrigation.

Wirtz contacted Gerald W. Peck, president of the Chicago firm, and convinced him that the potential for building a series of dams on the Colorado was well worth investigating. Peck came to Texas, looked into the situation, and promptly began successful negotiations for water rights with the Alexander sons, Jay and C. H., Jr., with the Syndicate Power Company and with Ward S. Arnold, an associate of the Alexanders who had been an engineer for Middle West Utilities.

A new firm, called the Emery, Peck and Rockwood Development Company, was organized with the covert backing, it was disclosed later, of the Insulls. (The Insulls thus did not let all of that $500,000 they had spent earlier get away from them.) The Alexander estate came out with a twenty-five percent interest in that firm, which emerged in July 1929 with the six major water permits it needed to begin construction of a large dam at what later became the site of Inks Dam. Because that site would have inundated large lead and graphite deposits, the new firm decided to move two and one-half miles upstream — to precisely the spot General Johnson had marked in 1854.

The Emery, Peck and Rockwood Development Company also decided to use the same name for the dam that had been chosen earlier by the Insulls: Hamilton Dam. The name honored G. W. Hamilton, vice-president and chief engineer for the Insulls' Middle West Utilities Company and a longtime, ardent promoter of the project.

Emery, Peck and Rockwood kept its interest in the deal secret at first. Wirtz opened an office for the firm in the Scarbrough Building, still a downtown Austin landmark, under the name "Associated Engineers," in August of 1929. Ward Arnold was placed in charge of that operation, which had three other employees: Jay Alexander and Ray E. Summerrow, who worked as land agents, and Beverley Randolph, secretary. Miss Randolph had worked for Wirtz ever since his election to the Senate and, not content to sit at a desk, she quickly began helping the men negotiate for land rights needed to build Hamilton Dam. In a January 1985 interview, she recalled:

> We used to travel in an old, open Ford up to Llano or Burnet to negotiate with land-owners, and it was an all-day trip. But then we'd rent a plane from the Miller Flying Service — I believe it was an old Waco biplane. Reagan Dickard was our pilot and also our draftsman. We'd fly up there and land right in a field and that sure saved us a lot of time.

On March 12, 1930, the Fargo Engineering Company was hired to draw plans for the series of dams the Insulls planned. On April 1, 1930, the Associated Engineers office began operating as the Emery, Peck and Rockwood Development Company. Quite a few more people were hired, and a year later everything seemed to be in place despite the demoralizing effects of the Great Depression.

Then, on April 18, 1931, an eight-column banner headline

across the top of the *Austin American*'s front page proclaimed: "Contract Awarded for Colorado River Power Dam." The article reported that a $3 million contract had been awarded on April 15 to the Fegles Construction Company, Ltd. of Minneapolis for a dam 9,000 feet long and 137 feet high. The structure was to be completed by March 1, 1933. Beverley Randolph said, "Everyone in our office had been expecting the news for some time, and we were really excited when it finally came. As I recall, we didn't do much work the rest of that day. And then, when the newspapers printed the story, it caused a lot of excitement — especially since it said the project would give work to hundreds of people."

The *Austin American* article said that Hamilton Dam would be operated by the Central Texas Hydro-Electric Company and that its total power output had been sold, although it did not name the purchasers. Later, it was revealed that they were the West Texas Utilities Company and the Central Power and Light Company — both owned by the Insulls. On November 28, 1930, they had signed thirty-year contracts with each company agreeing to purchase fifty percent of the electricity generated, at a rate of 7.5 mils per kilowatt-hour.

Meanwhile, the Central Texas Hydro-Electric Company had bought the Hamilton Dam properties and water permits from Emery, Peck and Rockwood Development Company for $2,876,500 in cash. Records in the Board of Water Engineers files revealed that the company also assumed all of the obligations involved except for $2,516,000 in notes.

Central Texas Hydro-Electric raised the money to buy the Emery, Peck and Rockwood assets by issuing $3,714,500 in bonds, paying six percent interest. Mississippi Valley Utilities bought $2,873,000 worth of these bonds, while Central and Southwest Utilities bought the other $841,500. Both were subsidiaries of the Insulls' financial empire.

Miss Randolph, who recorded in 1974 these and other events leading to the formation of the LCRA, wrote: "At that time, the Insulls controlled the largest utility company in the United States and their involvement in the Colorado River project seemed to assure its success."

She and other employees of Emery, Peck and Rockwood Development Company were transferred to the payroll of the new Central Texas Hydro-Electric Company. F. A. Dale, an engineer from Chicago, continued his role as manager for operations. Wirtz was placed in charge of all executive and policy matters.

Miss Randolph recalled that preliminary work on the Hamilton Dam had begun before the Fegles contract was announced and that the tempo picked up quickly. Fegles built houses for employees, a mess hall, temporary office quarters, and various other buildings. Soon the site "took on the look of a lively little village, teeming with activity."

Sim Gideon, who had gone to work as a young lawyer in the Wirtz firm at Seguin in 1930 and who was destined to become the third general manager of the LCRA, also recalled that busy scene.

> They built a construction camp out there in Llano and Burnet counties. I guess they had more than two thousand people out there working on the dam. They even had a hospital, their own cafeteria and, of course, places for the workers to live. There was quite a construction boom going on in that general area. And then, of course, the depression came along . . .

The Insull brothers, Gideon noted, had developed an interesting system for pyramiding their assets in the electric utility business.

> They created a holding company and it would control the voting stock of various electric utilities. Then, in order to build another project, they would have those electric utilities that were in operation make contracts with the project to be built — and then they would sell bonds based upon those contracts. But they would continue to control the company by keeping the voting stock.
>
> They had West Texas Utilities and Central Power and Light Company enter into a contract with the Central Texas Hydro-Electric Company, whereby they would buy power developed at Hamilton Dam. Those contracts were then pledged to secure bonds that were issued by Central Texas Hydro-Electric. The bonds were sold and the money from them was used to start construction of what was then called Hamilton Dam.
>
> But when the Depression came along, their pyramids all fell down — and you had quite a widespread feeling in the United States against the electric utility business as then operated and developed.

For the Central Texas area around Hamilton Dam, the boom turned to bust almost overnight. A warning signal was flashed early in 1932 but, at the time, none of the employees realized the severity of the Insulls' financial situation. Miss Randolph explained:

> We were notified that all employees would have to take a twenty percent cut in salary if they wished to stay with the company. This was to be effective February 1, 1932. While this news was not welcome, it was not unexpected and Mr. Dale advised the Chicago office that "the employees took the news of the cut in a very fine spirit."
>
> We knew that the budget had been cut because of the economic conditions and that we had been advised to keep expenditures to a minimum. However, land trades were still being made and other activities went on as usual.
>
> We should have been prepared — but of course we were not — when word was received on April 20, 1932, that the job was to be closed down that day. The blow had fallen and the unthinkable had happened. The Insull empire had collapsed and, with it, the Central Texas Hydro-Electric Company project.
>
> . . . At this time, the depression was really being felt in Texas as well as the rest of the nation.
>
> When the lights went out over the sprawling construction site on the night of April 20th, 1932, we did not realize that it would be over two years before we would ever see such a site again. It was such a shock that it was hard to realize that it had really happened. Many of the people working on the dam just could not believe that the job would not open up in a few days and they remained at the site. They didn't leave for some time.
>
> The closing down of the job also caused widespread concern among the people on the Colorado River who had backed the project and were looking forward to the benefits they would receive upon the completion of the dam.

An estimated $3,857,439 had been spent on the dam, which was approximately forty-five percent complete.

On April 25, 1932, the Fargo Engineering Company, engineers for the project, filed suit against the Central Texas Hydro-Electric Company in the U.S. District Court at San Antonio. Judge Robert J. McMillan appointed Senator Wirtz as receiver for the bankrupt company and gave him permission to hire Ray Summerrow and Miss Randolph as his assistants.

Central Texas Hydro-Electric owed $3,714,500 to bondholders, $751,000 to the Fegles Construction Company, $42,000 to Fargo Engineering, and smaller amounts to several other firms.

Samuel Insull, then seventy-two, fled to his native England in 1932 to avoid prosecution on fraud charges in the wake of the stock market collapse. But he was extradited in 1934 and brought back to Chicago, where he was tried — and acquitted — in three separate trials for fraud, violation of federal bankruptcy laws, and embezzlement. He went back to Europe and died on July 16, 1938, in Paris.

Martin Insull, still a British citizen, also was tried and acquitted but was deported to Canada. He died on May 4, 1947, in a nursing home at Orillia, Ontario, at the age of seventy-five.

It was during the fall of 1932 that Ralph W. Morrison, a San Antonio financier who also had made a fortune in the utility business, came up with a scheme to finance completion of the Hamilton Dam project. He found that the bankrupt Mississippi Valley Utilities Company owned $2,873,000 in Central Texas Hydro-Electric bonds and that U.S. District Judge Walter C. Lindley had named Eugene V. R. Thayer of Chicago as receiver for that firm.

Morrison went to see Thayer and, armed with the knowledge that those bonds were inferior to approximately $1 million in material and mechanics liens, persuaded him to make a deal. Thayer agreed that his best bet for recovering any of that money would be to trade the bonds for forty-nine percent of the common stock in a firm Morrison would organize to finance the completion of the dam. The agreement was contingent on Morrison's being able to raise the necessary funds and upon approval from Judge Lindley, as well as approval by Wirtz and Judge McMillan.

Wirtz met with Morrison and Thayer in Washington, D.C., to discuss the matter. According to Miss Randolph, Morrison had many influential friends there (he reportedly made a $100,000 contribution to President Franklin D. Roosevelt's 1932 campaign) and felt he could obtain some funds from the federal government.

Morrison organized the Colorado River Company, with three of his employees as its officers: C. G. Malott, president; E. W. Staple, vice-president; and I. C. Boughton, secretary-treasurer. Although the firm's capital stock was only $1,000, it promptly applied to the Reconstruction Finance Corporation (RFC) for a $4.5 million, ten-year loan at five percent interest to complete Hamilton Dam.

E. N. Hurley and C. A. McCulloch, receivers for the Middle West Utilities Company, filed protests both with the RFC and with Judge Lindley. They threw cold water on the whole idea, complaining that West Texas Utilities and the Central Power and Light Company (both subsidiaries of Middle West) should not have to transfer their power purchase contracts to any other firm. Officials of those two companies said they had opposed the contracts initially but had been forced by their superiors in the Insull empire to make them.

Hurley and McCulloch also argued, in a brief filed with Judge Lindley, that if the idea was so great and the project could be financed entirely with borrowed money and without risk, Thayer should do it on his own.

"Why do Mr. Thayer's co-adventurers get fifty-one percent of the net profit of an enterprise to which they contribute not a penny?" they asked.

Eventually, the question of the controversial power purchase contracts was resolved when Judge Lindley approved cancellation of them in return for Central and Southwest Utilities' transfer of its $841,000 in bonds to the Mississippi Valley Utilities.

In the meantime, the RFC had turned down the Colorado River Company's loan application due to a policy it had adopted of denying loans for all power and irrigation projects. But another door to federal funds had been opened with the creation in 1933 of the Public Works Administration (PWA), which Congress had given $3.3 billion along with the authority to make loans and grants for worthwhile projects designed to stimulate construction and put people back to work.

To Wirtz, Morrison, and other advocates of the Hamilton Dam, that provision of the National Industrial Recovery Act appeared to be exactly what was needed to provide financing — especially since completion of the project could involve as many as 2,000 jobs.

Unfortunately, it did not appear that way to Secretary of the Interior Harold Ickes, who also served as public works administrator. A lawyer who had been a newspaper reporter in Chicago for several years, Ickes was a maverick Republican who had strongly supported President Roosevelt and who would become known as "Honest Harold" for his efforts to keep the PWA's projects free of politics and corruption.

Ickes noted that the Colorado River Company was a private corporation dedicated to making profits while the PWA was created to provide financial assistance for public agencies. He turned down the $4.5 million loan application during the fall of 1933 and, once again, the Hamilton Dam project was left high and dry.

Quarterback Wirtz was not willing to concede defeat, especially since U.S. Senators Tom Connally and Morris Sheppard, along with Congressmen Buchanan and Mansfield, were all eager to help carry the ball.

Buchanan became even more enthusiastic about the project after Wirtz persuaded the Texas legislature, in redrawing congressional districts, to make Burnet County a part of his Tenth District. That, of course, put the half-finished Hamilton Dam (which eventually would be renamed Buchanan Dam) in his district.

And while Wirtz did not realize it during those dreary days of 1933, he soon would have on his team the congressional equivalent of a Heisman Trophy winner: Lyndon Baines Johnson, then an ambitious and skillful secretary to Congressman Richard Kleberg.

But the first thing Wirtz had to do was score some points in the legislature — against determined and colorful opposition. That opposition was led, ironically, by Representative Sarah T. Hughes of Dallas, who was later appointed by President John F. Kennedy as a U.S. district judge. On November 24, 1963, she would administer the presidential oath of office to Lyndon B. Johnson.

5

A Fair
Trade

A host of zealous, unpaid lobbyists led interference in the Texas legislature for Wirtz and other professional LCRA proponents during 1933 and 1934. They included several prominent citizens of Burnet and two of those finally hit pay dirt in the Governor's Mansion.

District Judge Thomas C. Ferguson of Burnet went to bat for the project from the beginning and eventually served five different terms on the LCRA's board of directors. A precocious newspaperman-turned-lawyer, he became a distinguished jurist and a walking encyclopedia of LCRA history.

Born in Roswell, New Mexico, he was twelve years old when his family moved to Burnet in 1918. Electricity was scarce and most people had no idea what a dam producing hydroelectric power could do for them. Judge Ferguson recalled during an interview on April 12, 1985:

> Mr. W. C. Galloway was a banker here in Burnet then. He had a "one-lung" diesel — a pretty good-sized diesel — and a generator. They'd start it up about sundown every day and run it until about ten o'clock, then shut it down. That was our only source of electricity. They'd start it up again the next morning

39

and run it until after breakfast — except on Tuesday mornings, when they'd run it until ten o'clock so the ladies could iron. Monday was washday so Tuesday was the day they ironed.

That lasted for three or four years after we moved here and then they started running it fulltime. Nobody thought about having much in the way of electric motors. Mostly, it was just lights. Then, in about 1927, the Texas Power and Light Company ran a transmission line into here and one into Marble Falls — but they didn't go to Johnson City.

That might have been a tactical error of classic proportions in the private vs. public power war, in view of later developments.

Ferguson was only fourteen years old when he graduated from Burnet High School in 1921. At fifteen, he resurrected a weekly newspaper, *The Index,* at nearby Liberty Hill; he sold it the following year, then served as editor of the *Blanco Courier* in 1923 and 1924 and as publisher of the *Burnet Bulletin* in 1925 and 1926. After a short stint on a small-town newspaper in Arizona, he returned to Burnet in 1927 and became deputy district clerk. He quickly learned enough law to pass the bar exam, and he became a lawyer in December 1928.

R. E. Johnson, the son of Gen. Adam Johnson (no relation to LBJ), was district clerk and also had an abstract business. When the Insulls became interested in building Hamilton Dam, they contacted Johnson to examine abstracts. He farmed out much of the work to young Ferguson.

By the time the first LCRA bill showed up, Ferguson knew the subject well — and he was not exactly a political amateur. He had served as Burnet County Democratic Party chairman in 1929 and 1930, then one term (1931–32) in the Texas House of Representatives, losing his bid for reelection. Later, he served as mayor of Burnet and as Burnet county judge before becoming a district judge in 1947.

The half-finished Hamilton Dam stood as a grim reminder to Ferguson and other civic leaders of what the construction jobs on the project had meant and could mean again to the local economy.

"We were anxious to see the dam completed because of the jobs," said Ferguson, "and we also felt it would be a tourist attraction. It turned out to be more of a power deal than we had visualized."

Meanwhile, the Wirtz forces in general — and Wirtz in particular — needed all the help they could get.

Governor "Ma" Ferguson took office for the second time on January 17, 1933 (her first term ran from January 20, 1925, until January 17, 1927). She convened the first called session of the Forty-third Legislature on September 14, 1933, primarily to pass state laws needed to qualify for much of the aid proffered by President Roosevelt's New Deal legislation.

Wirtz wrote a bill designed to create a Colorado River Authority and got it introduced on October 4. The measure won quick passage in the Senate but did not come up for a vote in the House before the session adjourned on October 13. The second called session convened on January 29, 1934, but the Colorado River Authority bill was not introduced until February 20; once again, it passed the Senate but the session ended on February 27 without any action being taken in the House.

In Washington, meanwhile, Congressman Buchanan adamantly refused to take "no" for an answer on the long-sought $4.5 million loan. His efforts drew strong support from Texas Senators Connally and Sheppard. On February 9, 1934, Buchanan informed the Colorado River Improvement Association that the loan application was on Secretary Ickes' desk and declared: "I argued the case before the President in the most intelligent way I could and he was deeply interested in it. I will advise you just as soon as I know what the final outcome will be."

In Seguin, Wirtz narrowly escaped death on Monday morning, February 26, 1934, when an irate landowner came calmly into his office and, after a brief conversation, suddenly started shooting. He killed Gerald Peck, the Chicago banker who was a founder of the Emery, Peck and Rockwood firm. In an article which appeared in the *Austin American* on February 27, 1934, Gordon Fulcher reported:

> SEGUIN, Feb. 26 — A difference of long standing over a land damage claim Monday night had resulted in a fatal shooting of G. W. Peck, power company official of Chicago, and a retired Seguin farmer's being charged with murder.
>
> Peck was shot to death by Tom Hollamon, 76, in the office of Former Sen. A. J. Wirtz, one of the officials of the company and a close personal friend of Peck.
>
> Peck, Wirtz and F. S. Hunt of Jackson, Mich., were in the of-

fice at the time waiting for time for the annual meeting of the Texas Power Company stockholders Monday morning. Hollamon fired two shots, the first going wild and the bullet embedding itself in a wall. The second struck Peck about six or seven inches below the right armpit and left the body slightly below the left armpit . . .

Fulcher said Hollamon was released on $12,000 bond after an examining trial Monday afternoon at which he refused to testify. He added:

> Hollamon appeared at the examining trial with his head bandaged from an abrasion received when he and Wirtz grappled for the .41 caliber single action Colt pistol which fired the fatal shot into Peck's body.
>
> Wirtz received a slight cut on the side of his head in the struggle. Hollamon was hurt, Sheriff Albert W. Saegert said, when he struck his head against the door in the scuffle.
>
> The piece of land over which the fatal shooting occurred is located on the banks of a lake formed by one of the power company's dams there. Hollamon claimed that backwaters from the lake had ruined his land. He already had been paid one judgment of $3,000 by the power company. Dissatisfied with the judgment, several months ago he filed another suit claiming additional damages and a survey to determine the extent of damage to the land was being conducted by the power company.

Hunt testified at the examining trial that Hollamon came into the office about 9:35 A.M. He, Wirtz, and Peck all shook hands with the white-haired farmer, who had once served as a U.S. deputy marshal. He said Hollamon declined Wirtz's invitation to take off his overcoat and have a seat, saying he would only be a few minutes.

Hollamon said he had come to see them about settling his claim, declaring there had been too much delay and he wanted to know when the survey would be completed. Wirtz and Peck both assured him it would be handled as quickly as possible, according to Hunt.

"There was no argument on the part of Mr. Peck or Mr. Wirtz and I took no part in the conversation," said Hunt.

Hollamon again declared he was not satisfied and started to open the door but then turned suddenly and declared, "I came in here to get a settlement and I am going to have it!" With that, he pulled the pistol and fired the shots, according to Hunt.

Peck slumped to the floor immediately and Wirtz yelled: "No, Mister Tom, no!"

Wirtz then jumped up and grappled with Hollamon, grabbing his gun hand as they fell into the hallway. Hunt and F. G. Chamberlain, a San Antonio engineer who had been outside, joined the brief struggle.

"You can have it," Hollamon said as he released his grip on the pistol, Hunt recalled.

Sim Gideon was in an office across the hall when the shooting occurred.

"We were right above a little bank and I thought at first the bank was being robbed," Gideon said during an interview at his Austin home on March 7, 1985. "I probably should have locked my door but, instead, I ran out there and took the gun, then put it in my desk."

Someone found a pillow and put it under Peck's head as he lay in the hallway. He died about thirty minutes later.

Rudoph A. Weinert, Wirtz's law partner and also district attorney, was out of town at the time. When he returned late that afternoon, he spotted the pillow before he found out what had happened. In his typically gruff manner, he growled: "What the hell? People taking naps up here and bringing in their pillows?"

According to Gideon, the murder case was transferred to Lockhart and Hollamon died about a year later, before it came to trial.

A short time later, Wirtz moved to Austin. Other sources said it was primarily because Hollamon's bitterness was shared by many of the senator's former constituents, who resented his role in eminent domain matters involving the Guadalupe Valley dams, and that he was told to "get out of town." Gideon, who followed Wirtz to Austin in July, and Miss Randolph, who also worked for Wirtz at the time, both belittled that theory; however, both admitted to being unabashed admirers of the man most of his friends referred to as "Senator." Not "*the* Senator" but just "Senator."

Wirtz, Judge Ben H. Powell, John Rauhut, and Gideon organized their law firm shortly after the Peck shooting and it became highly successful. Rauhut and Gideon had been University of Texas Law School classmates, graduating in 1929. Rauhut had gone to work immediately for Judge Powell while Gideon spent a year with the law firm of Church and Graves in San Antonio before joining Wirtz.

"It was hard to leave them, too," Gideon recalled with a chuckle, "because our office was in the Milam Building — and it was the first air-conditioned office building in Texas."

But a court reporter who also had an office in the Milam Building heard that Wirtz was looking for a young lawyer and he recommended Gideon.

"I guess I just happened to be in the right place at the right time," said Gideon.

That was not the last time. When he moved to Austin in July 1934, he lived in a house at No. 1 Happy Hollow Lane. Lyndon Johnson married Claudia Taylor on November 17, 1934, and they moved into a house across the street, No. 2 Happy Hollow Lane.

"We were friendly from then on," Gideon said of LBJ, "for all the time he was in Congress and while he was vice-president and then president. We were fortunate enough to be invited to dinner at the White House and things like that. It was a big thing for a little boy from Coleman to accidentally be thrown into that."

Gideon recalls that one of the many things Wirtz and Powell had in common was an intense dislike for Governor Dan Moody, who became the state's youngest governor when he took office in 1927 at the age of thirty-three. The feeling was mutual, said Gideon, and the bitterness of the Moody supporters toward Wirtz probably created some of the widespread criticism of him.

Powell had served on the Commission of Appeals, a six-member arm of the Texas Supreme Court created by the legislature to relieve the high court's overload of cases. At that time, the state constitution provided for only three members of the Supreme Court. Powell had been appointed by Ma Ferguson and he became angry when Moody refused to reappoint him, according to Gideon.

"Wirtz fell out with Moody while he was in the Senate," said Gideon. "Senators sometimes think a governor doesn't do what he promised to do. So they had a mutual dislike of Moody and he of them . . . anyone who talked with any of Moody's friends would get an entirely different idea of Wirtz than he'd get if he talked to Wirtz's friends."

Meanwhile, Wirtz's friends in Washington continued their efforts to tap federal funds for completion of the Hamilton Dam. Until June 28, 1934, the proponents of the project had no assurance that they could get any federal money even if a state agency were created to handle it. But that night, the Austin Chamber of Commerce held its annual banquet with Senator Tom Connally as the main speaker.

Connally began his remarks by reading a telegram from Congressman Buchanan which said President Roosevelt had just approved, conditionally, a Public Works Administration loan of $4.5 million for the Hamilton Dam project. The president was quoted as saying: "It would be worth $4,500,000 to stop those floods and conserve the water if no other use were made of the dam."

That certainly drew no argument from the banquet audience, which greeted the news enthusiastically despite the main condition: that it be made a public project instead of a profit venture for the Colorado River Company. That firm would also have to sell all the property involved at a "prudent investment cost" to a state agency if the legislature should create one to take over the project.

Victory seemed close enough that leaders of the Chamber of Commerce and the Colorado River Improvement Association immediately began planning an elaborate celebration, primarily to honor Buchanan for his successful efforts. Wirtz and Mansfield also were honored at the huge party, which was held at the Hamilton Dam site on July 17, 1934.

Buchanan urged speedy passage by the legislature of a bill to create a Colorado River Authority. He also promised that if such an agency were established, the federal government would not only help complete Hamilton Dam but also would help finance four more dams between Burnet and Austin.

"This will be the biggest thing, next to the Tennessee Valley Authority, in the nation," he declared. "Llano, Burnet and Travis Counties will become the hydro-electric power center of Texas."

His audience included 297 citizens representing twenty counties and twenty-six towns in the Colorado Valley. To show their appreciation for the congressman, they voted unanimously to change the name of the project to "Buchanan Dam" and ask the legislature to switch Burnet County into his district.

Wirtz already had begun steps to terminate the receivership of the Central Texas Hydro-Electric Company and thus clear the decks for a takeover by a public agency. By October 5, that problem, including the sale of assets and cancellation of those troublesome power-purchase contracts with Insull subsidiaries, had been settled. The Colorado River Company emerged with the Hamilton Dam property, six valuable water permits, and twenty-eight parcels of land — all of which it would be happy to sell to the proposed public agency.

But in the meantime, the third called session of the Forty-third Legislature had refused to pass the LCRA bill. The defeat came in a photo finish. The session convened on August 27, 1934, and the bill was introduced on August 30. Two days later, a joint committee of the House and Senate heard testimony supporting it from Wirtz, Buchanan, Mansfield, and the chief counsel of the PWA, Henry T. Hunt.

Mansfield said he favored the measure because the proposed series of dams would protect the valley below Austin from ninety percent of the flood damages it would suffer otherwise.

"I know of no other river in the United States, except the Mississippi, which has flood damage exceeding that of the Colorado and the Brazos," he declared.

The Senate passed the bill on September 17, by a vote of 25 to 1. That night, the House Committee on Conservation and Reclamation held a dramatic hearing on it.

Representative Harry McKee of Port Arthur moved successfully to substitute the Senate bill for the measure which had been introduced in the House. Representative Fritz Engelhard then explained the bill, according to the official minutes recorded by Bess Elkins, and declared:

> . . . this is the greatest water proposition which has been presented in the southwest. This prospective dam will store more water than Elephant Butte and the Boulder Dam. It will be the largest dam in the southwest and will impound one million, one hundred twenty-five thousand acre-feet and will create a lake 187 miles long in the Colorado River . . .

Wirtz then testified, declaring that he did "not represent any interest that is getting one cent out of this bill." He added that he was acting as receiver for the Central Texas Hydro-Electric Company and had tried to refinance the Hamilton Dam project which it had started. Said Wirtz:

> There has come to me, rumors that Mr. Ralph W. Morrison's connection with this will make him a lot of money. As far as these rumors are concerned, when I hear them that come from

legislators I do not doubt their authenticity, because I do not doubt the word of any man in this legislature, but I do want to point out to you that these rumors were disseminated before this legislature came into session.

There is a man in Chicago, Mr. Chesley Manley, working on the *Chicago Tribune,* who has been casting aspersions and causing all these rumors. I have a file of all the clippings, which I will be glad to let you see. Mr. Manley attacked every project, every power project, to which the PWA had lent assistance in the United States — beginning with the Grand Coulee project and ending with this one.

He tried to make three things clear. I have heard them here: first, that these loans were made as a result of political persuasion. Second, that they would never be paid off, and third, that there was too much power being generated now and that the government ought not to be in competition with private companies. Mr. Morrison is said to have made huge donations to the Democratic National Committee. I doubt that not — but it has nothing to do with this project under discussion . . .

All that the PWA asks is that you charter a public corporation under public control with the right to go out and harness this stream and take advantage of his Public Works money . . .

Three days later, the committee approved the bill.

When it came up for floor debate in the House, a group of opponents led by Sarah T. Hughes presented twenty-seven proposed amendments. Two major ones were adopted: (1) by Rep. W. V. Dean of San Saba, making all rights of the new facility subordinate to rights of cities and towns within the entire watershed, and (2) by Hughes, prohibiting the agency from paying any money to R. W. Morrison, or to his Colorado River Company, until after the anticipated federal government loan had been repaid.

Few of the representatives were aware of Morrison or his involvement until Mrs. Hughes "explained it" to them. According to the *Dallas Morning News* of September 23, 1934, she said:

> When this property was sold two weeks ago it was bought by the Colorado River Company and the Colorado River Company is composed of an employee of Mr. Morrison. Now let's get to the bottom of this thing. Who is Mr. Morrison?
>
> Mr. Morrison is a public utility man. He owns the largest public utility in Mexico. He owns the Laredo Power and Light Company. He was part owner of the Insull properties in Texas.

Mr. Morrison was the largest contributor to the Democratic campaign from Texas. He was the second largest contributor to the Roosevelt campaign in the United States.

As a reward for that contribution, Mr. Morrison was appointed to the disarmament conference. As a further reward for his $100,000, he sought to get a loan to complete the Hamilton Dam. He went to Secretary Ickes and was refused. He went to see the President then, and the President authorized the loan if we would create an authority.

Now Mr. Morrison's reward is to be 51 percent of the stock of the Colorado River Company, while 49 percent goes to the owners. That is a pretty good commission for getting this loan. He wants to get his money at the same time the United States gets its money. Mr. Hunt, as attorney for the PWA, told us that $3,200,000 is a reasonable value of the property and Mr. Morrison would get 51 percent of that. I would not want that tainted stock. If I were a multimillionaire, I'd give this to my State and my Nation.

The House finally passed the bill, 95 to 14, but the Senate refused to accept the House amendments and the measure went to a joint conference committee.

At a dramatic meeting of that committee on the last night of the session, September 25, the five Senate conferees and two of the House conferees voted to eliminate both the Dean (water rights) and the Hughes (Morrison money) amendments.

At least three members of each five-member delegation on the conference committee would have to sign the report before it could be considered by the House and the Senate. Representative Olan R. Van Zandt of Tioga, who was blind, had his pen poised and was ready to sign when Mrs. Hughes spoke up: "I'm not going to sign this. You aren't, are you, Mister Van Zandt?" Then she turned to Representative Will H. Scott of Sweetwater. "You aren't, are you, Mister Scott?"

She asked both Van Zandt and Scott to leave the room with her and thus kill the bill. Although the other conference committee members urged the two men to let the House vote on the measure, they joined Mrs. Hughes and marched out, letting the bill die a natural death when the clock struck midnight.

Once again, Wirtz obviously needed a little help from his friends. And two of his best ones, according to Gideon, were the Governors Ferguson.

"I was close to the LCRA proposal before the bill was passed," said Gideon, "although I was too young to realize what was really going on. But I was aware that Senator Wirtz was a big friend and supporter of the Fergusons. Jim Ferguson said he was going to get that bill passed — and finally, he did."

Jim Ferguson officially became governor of Texas on January 19, 1915. During his first term, Ferguson received generally high marks for making badly needed improvements in public education, especially in providing free textbooks for public schools and for winning passage of a compulsory attendance law. Ironically, higher education was at the root of his downfall. He was reelected to another two-year term in 1916 but was locked even then in a bitter fight with The University of Texas Board of Regents. That struggle is widely credited with leading to his impeachment in 1917.

Ferguson was charged with misuse of state funds, with having state funds deposited in a bank in which he owned stock, and with receiving a $156,000 loan, that was never repaid, from the Texas Brewers Association in 1914.

The House of Representatives voted twenty-one articles of impeachment against Ferguson and the Senate upheld ten of them, removing him from office and prohibiting him from holding public office in Texas again. But he had anticipated the verdict and resigned a short time before it was returned, contending later that this nullified the office-holding prohibition.

He tried unsuccessfully to get the next legislature to rescind that ban. When he failed, he masterminded his wife's successful campaigns in 1924 and 1932, telling voters they could get "two governors for the price of one."

The voters did, although there were some cynics who came to doubt that "Ma" was carrying her share of the load in the governor's office. That theory was reinforced by the fact that his desk was next to hers in the executive office, and she usually left about midafternoon while he stayed until around 6:00 P.M.

It was "Governor Jim" who greeted a small delegation from Burnet, headed by Christian Dorbandt, during the fall of 1934 and made a pledge that paved the road to Colorado River development. Dorbandt, a former Burnet County sheriff who owned a large ranch fronting on the river, was a close friend of Jim Ferguson, according to Judge Tom Ferguson (who was not related to the governors). Dorbandt made a plea for gubernatorial help in getting the LCRA bill passed. Judge Ferguson explained the circumstances:

Everybody in Burnet was affected tremendously by the closing of the project out there. Being in a depression, too, everybody up here wanted it and we were all helping the people downriver from Austin who had been trying to get flood control. When the Brownwood Chamber of Commerce voiced their opposition to the dam, the Burnet Chamber of Commerce and the people downriver got together to promote it. We had lots of help.

Dallas was greatly opposed to it — I think largely because of the Texas Power and Light Company, which had distribution rights all through this area. Mrs. Hughes was a member of the legislature and she was very much opposed to it, as were all the members from Dallas.

But the state was making plans at the time for the 1936 Centennial, celebrating the 100th anniversary of Texas's independence, and Dallas had won a tough fight to be designated the headquarters city for the festivities. The Dallas delegation was trying to get a bill passed that would help finance the big party by giving the City of Dallas $5 million in state funds, with that amount to be matched by a federal grant. Said Judge Ferguson,

That bill hadn't passed in earlier sessions, and the Dallas people were just as interested in it as we were in the LCRA bill. So Jim called the Dallas legislators into the Governor's Office and told them, "I'm going to call another special session and there are going to be two bills submitted — the centennial bill for Dallas and the LCRA bill. And you ain't gonna get one without the other!"

That was his idea — tying them together. He knew how bad they wanted that centennial bill.

Judge Ferguson was a close friend of Representative George Parkhouse of Dallas, having helped him get a job as assistant reading clerk in the House before Parkhouse was elected to the legislature.

But Parkhouse had been voting against the bill and Ferguson "couldn't do anything with him." Then, the judge recalled, "on the first day of the fourth called session, I was waiting out in the House Reception Room. He came to the door and I started for him. He threw up a hand and said, 'I'm for you, I'm for you!' "

The date was October 12, 1934. Judge Ferguson realized then that success finally was within reach.

6

Big
Bargain

Governor "Ma" Ferguson's official proclamation convening the
Forty-third Legislature's fourth called session at noon on October
12, 1934, listed five topics which could be considered. But most of
the controversy surrounded the LCRA bill and the proposed appropriation for the Texas Centennial. The other topics included the
approval of an additional $3.5 million in state relief bonds ("bread
bonds," as they were called), remission of interest and penalties on
delinquent taxes, and the remission of taxes for the Brazos River
District to secure public works loans "for improvement and work-
creating" projects.

   All these matters had been considered initially at the regular
session in 1933. And the Colorado River Authority measure also
had been considered at the three preceding special sessions. This
time, the bloodletting within the legislative chambers was backed
up by a heavy bombardment of newspaper editorials plus urgent
appeals to Washington. And the situation was complicated to some
extent by the fact that the Fergusons, plagued by state financial
problems, decided to withdraw from politics; Ma did not run for reelection in 1934 and, during that fourth special session, Attorney
General James V. Allred was elected to succeed her.

Ironically, the general election and that fateful special session frequently shared front-page space with the lengthy, dramatic mail fraud trial of Samuel Insull and sixteen other defendants, which was taking place in Chicago.

Allred won the general election on November 6, four days before the mandatory end of the thirty-day special session. By then, the bill appropriating $5 million for the centennial — and levying $5 million in taxes on corporations to pay for it — was dead. And the LCRA measure was gasping for breath as its opponents prepared to administer the last rites.

Both presiding officers, Lieutenant Governor Edgar Witt and House Speaker Coke R. Stevenson, had gone to bat for the LCRA bill. Witt even signed as one of the cosponsors, an unusual move for a lieutenant governor, when Senator John Hornsby of Austin introduced it. Other cosponsors included Senators T. J. Holbrook of Galveston, Welly K. Hopkins of Seguin, Walter Woodward of Coleman, Archer Parr of Benavides, Albert Stone of Brenham, Arthur Duggan of Lamb County, H. C. Woodruff of Decatur, W. R. Poage of Waco, and Will R. Martin of Hillsboro. An identical bill was introduced the same day in the House by Representative Fritz Engelhard of Eagle Lake, a rice farmer and longtime crusader for Colorado River development. Representatives Harry N. Graves of Georgetown and T. H. McGregor of Austin signed on as cosponsors.

The bill breezed through the Senate, 24-0, on October 17, 1934. That same day, the *San Angelo Evening Standard* published an editorial headlined "Alienating West Texas" and declaring:

> The special session's reception of the Colorado River bill thus far seems to indicate clearly enough a determination to do or die for the district, ignoring the protest that has gone up from this section over water rights . . .
>
> West Texas has heard much about a series of dams along its streams in recent election campaigns, has seen projects promoted under the relief program. If the proposed Colorado hydro-electric dam in no way puts a hurdle in the way of these projects, well and good; otherwise, the State of West Texas, such as Max Bentley proposed in a recent article in *West Texas Today,* may be forced as a means of retaining that which was placed naturally as its own. The old propensity to give West Texas away to the railroads, to the colonizers, to the schools, and now to the power developers, continues unabated.

Three days later, the West Texas Chamber of Commerce made a direct appeal to President Roosevelt for help in amending the bill to provide water priorities for "future domestic, municipal and irrigation users." James D. Hamlin, president of the organization, sent the following telegram to the president on October 20, 1934:

> This organization representing two-thirds of the territory of Texas and about three-fifths of the watershed of Colorado River is endeavoring to secure an amendment to the Colorado River Authority bill now before the Texas Legislature which will subordinate hydro-electric users of Colorado flood water to future domestic, municipal and irrigation users STOP Proponents of bill at Austin claim that bill is in accordance with your suggestion and also claim that Public Works Administration will not approve the federal loan to this project if bill is amended as above indicated STOP Unless amendment is secured it is our belief that an almost complete monopoly will be given to one user of these flood waters for power purposes and that many thousands of citizens, farmers and ranchmen will in the future forever be deprived of use of flood waters of Colorado thereby stagnating and blighting the future of a tremendous agricultural livestock and range country STOP Of approximately 1,400,000 acre feet of the flood waters the Colorado authority will have permits for about 1,100,000 acre feet and the balance is already impounded for other appropriations for domestic and irrigation purposes which results in drying up the Colorado for future use by five-sixths of the potential appropriators residing above proposed dam STOP We hope that on approval of Public Works Administration the legislature will amend bill as above indicated thereby making possible the building of project as well as protecting the water rights of ranchmen farmers and domestics whom we believe should have prior consideration STOP We pray for your earnest consideration and immediate attention of this matter as bill is now before legislature and in all probability will pass without amendment unless approval of amendment is immediately given by Public Works Administration STOP

Henry T. Hunt, general counsel for the PWA, replied to Hamlin with the following telegram on October 22:

> Retel Twentieth Colorado River Docket 380 STOP Your proposed amendment would jeopardize project by making possible reduction of waters available to amount insufficient for uses of

authority STOP Bill subordinates water rights to be acquired by
authority to other existing rights, thereby placing latter in better
position than if bill not enacted STOP Understand Senator
Woodward satisfied with provision which is necessarily a compromise STOP

Senator Woodward may have been satisfied with the assurance referred to by Hunt, but other West Texans were not. They in-
tensified their efforts to get the water rights amendment adopted.

They finally succeeded in the House, where a debate which
had been scattered over five days came to a climax on October 30.
Fourteen amendments were adopted, including one by Representatives Dean and Penrose Metcalfe declaring that the rights of the
Buchanan Dam reservoir would forever be subordinate to the municipal and irrigation requirements of the area above the dam. Another, also a sticking point, was one by Representative Weaver
Moore of Houston; it was similar to the one championed earlier by
Sarah T. Hughes and was designed to keep R. W. Morrison from
making any profit on the proposed project.

The bill won final passage in the House on a 107-to-6 vote and
went back to the Senate, which promptly refused to accept the
amendments and asked for a conference committee.

Meanwhile, former Governor Jim Ferguson made a strong
fight to revive the centennial appropriation and tax measure. The
Dallas Morning News reported on October 30:

AUSTIN, Texas, Oct. 29 — James E. Ferguson, a silent lis-
tener for two hours to pleas by utility companies against taxes to
support the Texas Centennial, went dramatically to the defense of
the Centennial and the proposed taxing program Monday night
before the House Committee on Revenue and Taxation. The
meeting was attended by the largest group of lawyers and executives of Texas business and industrial institutions that has assem-
bled in Austin in many months.

"Some of you say you are for a Centennial," Ferguson said,
"but you add, 'Don't tax me.' Come on, boys. Get in the game or
frankly say you are against it. Take a broad view of this.

"I hear it said we have a drought and a depression. That's
one reason I'm for the Centennial. We're going to need it. It will
do us a lot of good and we have a good, patriotic reason for holding it. Boys, kick in and go to shouting. Stop talking about going
to help Dallas and not help some other place. Call on your prin-
cipals back home and I'll guarantee you all of them will be broad-

minded enough to see the wisdom of this thing. Let's not be niggardly with our patriotic obligations.

"Call off your dogs. You can beat this bill. We all know that, but are you going to do it? Get this straight — if the Centennial fails it will be because the corporations will have caused it to fail. Do it for your own sake. The people can live over whether we have a Centennial or not. If anyone is left out that should be in, put him in. Open the subscription books and invite all to subscribe.

"I don't believe there is a man chopping cord wood in Texas for 75 cents a cord who wouldn't donate 75 cents as a subscription. Get in touch with the heads of your business and don't stop the Centennial."

Perhaps they should have contacted the wood-choppers; the business firms rushed headlong away from the opportunity to pay the increased taxes for centennial purposes.

And, once again, it seemed that about the only thing the conference committee on the LCRA bill could agree on was to disagree.

The Colorado River Valley Area Committee, which had been organized to recruit volunteer lobbyists for the measure and to coordinate their efforts, made an appeal for approval of the LCRA bill in a statement sent to all House members on October 22. It said the measure would permit completion of Buchanan Dam to begin within a short time and would furnish employment for at least 500 men a day for practically two years.

"It will give a number of honorable Texans an opportunity to earn a reasonable wage instead of depending upon Relief Funds for existence," the statement declared, then continued:

> If the bill in question is passed, the State of Texas will have an opportunity to secure without cost to the State or any of its citizens, flood protection on 300,000 acres of the richest valley land in the State and flood prevention to every county seat below Austin; and in addition, the dam will furnish a regulated supply of water that will make practicable the rehabilitation for the now-unused dam at Austin and the generation of probably $200,000 worth of hydro-electric energy per year, over that dam. In addition to that, farmers on the lower reaches of the Colorado River who have for the past years had to depend upon the unregulated supply and frequently wholly inadequate supply of water for taking care of their investment of more than one million dollars in ir-

rigation equipment, will have available a dependable supply of water that will make their business a successful one instead of a purely haphazard thing. In addition to this, the dam, when completed, will save many lives and prevent at least four million dollars worth of property damage annually.

We are advised that certain members of the Legislature are opposed to the bill in question because they say that, if it is passed, a large bonus and commission will be paid out of the stock of the Colorado River Co. We submit that no Texan can afford to oppose the bill on that ground.

In the first place, the Hamilton, or Buchanan Dam, now belongs to the Colorado River Co. All of its stock is owned by Eugene V. R. Thayer, Receiver of the defunct Mississippi Valley Utilities Investment Co. Thayer is not a representative of any public utility. He is the representative of the creditors and stockholders of the defunct corporation for which he was appointed receiver. While the stock of the Colorado River Co., with exception perhaps of some five qualifying shares, is vested in him, yet he does not own the stock, but owns it in his capacity as receiver for the unfortunate investors in the defunct Insull institution — the Mississippi Valley Utilities Investment Co. Thayer was authorized by the Federal Court appointing him Receiver to convey the property now belonging to the Colorado River Company to it in consideration of its capital stock.

If any of the creditors and stockholders of the Mississippi Valley Utilities Investment Co. realize anything out of the wreck, they must get it out of the stock of the Colorado River Company. The Federal Court in Chicago has absolute control over that matter. The Receiver can do nothing without the consent of the Court, by an order entered upon the Minutes of the Court. We have always been of the opinion that the Federal Courts were amply able to protect themselves and particularly amply able and anxious to protect creditors of the corporations for which they have appointed receivers.

If the Federal Court in Chicago authorizes Thayer to give the trust property away, is there any way this Legislature can prevent it?

If the CVA bill is passed and the authority created, it expects to acquire the Hamilton or Buchanan Dam. It does not expect to acquire the stock of the Colorado River Company. It will have nothing to do with that stock. Not only that, but the PWA has set limitations upon the price that the public authority shall pay for the property. It is provided that it shall not pay more than the prudent investment cost. There is no doubt about that. Mr.

Hunt, Chief Counsel of the PWA, made that statement to the Legislature.

This cuts out and prohibits the public authority from paying any bonuses or commissions. Cannot the Federal Government be trusted to take care of itself? Have you any reason to believe that the public authority which you are asked to create, when it purchases the property will pay any more than the limit set by the Government? Is it not more reasonable to believe that it will buy the property just as cheaply as possible?

If you pass the bill in question, this great enterprise will become a public enterprise and cost the State of Texas and its citizens not one penny. If you defeat the bill, it will remain in private hands. Not only will it remain in private hands, but you deprive at least five hundred citizens of Texas of jobs for two years at least . . . You will deprive those people of an opportunity to make a reasonable wage and get off the charity rolls. You will deprive every County Seat from Austin to the mouth of the Colorado River of flood protection. You will deprive the City of Austin of an opportunity to realize something out of its dam, which has cost it, all told, more than a million dollars. You will deprive the rice growers on the lower Colorado River of an opportunity to raise an assured crop annually. You will endanger the lives of many citizens of this State and you will sanction, by failing to prevent, the destruction of four million dollars worth of property annually from the Buchanan Dam to the mouth of the Colorado River. Its defeat benefits nobody but the power companies.

Now, as patriotic citizens, men who love your State, can you afford to deprive your State of those advantages in order to act as Guardian of a Federal Court, in a foreign jurisdiction, with the hope of preventing somebody from getting something from a Federal Receiver under the sanction of a Federal Court? We think the answer to this question, when you cast your votes on this bill, will be "No" — distinctly and audibly, "No"; that you will take advantage of the present opportunity of securing one of the most important enterprises that Texas has had an opportunity to acquire, and thereby protect not only the property of the citizens of the State, but their lives as well.

If this bill is passed, the advantages above enumerated will accrue to the State and its citizens, and no existing rights to water on the Colorado River or its tributaries will be affected by the passage of the bill. Their rights are not only protected by the bill itself, but by the law now on the statute books. Each one of you knows that no bill passed by the legislature can deprive anyone of vested rights.

The statement was signed, "Respectfully submitted, Colorado River Valley Area Committee," with the notation underneath that it had been read and approved by Engelhard, McGregor, and Graves, the House sponsors of the measure.

Congressman Buchanan, who had been lobbying in Austin throughout the session, may have had a hand in expressing those strong sentiments; but he knew that statement was not enough to turn the tide. He made a quick trip to Washington and urged the PWA to go along with the Dean water rights amendment. The following telegram, which he sent to Senator Woodward, has been credited with providing the final impetus for the passage of the bill creating the LCRA:

> I conferred at length with attorneys Hunt and Burke of the PWA on the Colorado River Authority bill now pending in the Texas Legislature. As you know I favor the development of Texas and every section thereof, therefore I want the Colorado bill to deal fairly and justly with the people throughout the watershed.
>
> Believing as I do that abundant flood waters go down this river which if conserved will meet every demand of municipalities, irrigation, and production of hydro-electric power, and further believing that I can procure the allotment of the necessary funds to complete the Buchanan Dam as well as three or four other dams on other good dam sites immediately below the Buchanan Dam with the Dean or public policy amendment in the bill, therefore neither public works or myself deem Dean Amendment fatal to the project.
>
> The Moore Amendment dealing with compensation, commission, etc., arising out of a contract entered into between receiver Thayer and R. W. Morrison about two and one-half years ago making it a felony for the court receiver and Morrison to carry out this contract, etc., is fatal to the project. It will render it impossible for the public authority to procure title and possession of the dam site and other property now vested in the Colorado River Company.

Buchanan's telegram undoubtedly played a major, if not the decisive, role in the conference committee's decision to accept the Dean amendment and reject the Moore amendment.

The Senate adopted the conference committee report by a vote of 27 to 1 on Friday, November 9, 1934, and the House scheduled a final vote on it for the following morning — with the session due to end that day at noon. A preliminary House vote had given it an 81-

54 endorsement; however, 100 affirmative votes (a two-thirds majority) was needed to put it into immediate effect.

On that Saturday morning, the *Austin American* had a front-page banner story on the Senate's action and the anticipated final passage of the bill by the House. One of the headline "decks" on that article noted: "Leaders in House Seek Two-Thirds Vote for Passage." Four columns to the left, also at the top of the front page, was a story headed: "Centennial Bills Definitely Lost in Closing Rush." And in between was a front-page editorial urging the House to give the "Colorado Valley Authority" the 100 votes needed to make the bill effective immediately. The headline said: "100 CVA Votes Will Mean Jobs for 1500 Needy Now." A picture of Buchanan was set into the editorial, which led the article with a caption:

> *Cong. J. P. Buchanan telegraphed Friday from Washington that after seeing President Roosevelt, he believed 15 million dollars ultimately would be spent for labor and materials for development of the Colorado River provided the Texas Legislature acts properly.*
>
> The Texas Senate 27 to 1 said Friday, "Certainly we want 15 million in government money to develop water power in Texas and put 1500 idle, hungry, worthy men to work promptly."
>
> The House said, 81 to 54: "We want no amendment which Cong. Buchanan says would prevent money coming from the national government which would block, possibly destroy, this project, or at least delay an employment program vital to hundreds of destitute families."
>
> Unless the improbable happens, the changing of minds of some 15 representatives while they sleep, the Colorado River dam bill will pass Saturday morning. Instead of passage by 81 to 54, is it not sensible that the vote in the house be a clear two-thirds majority, so that work may start promptly?
>
> Winter is here. Cong. Buchanan at Washington today is awaiting word of passage of this bill, so he may go at once to the proper Washington authorities with a request for immediate action. If the Texas Legislature does its duty today and if Gov. Ferguson immediately signs the bill, the federal government will be quickly put in motion sending millions to Texas. Trainloads of material will move up the river. Hundreds of men will be employed. Winter is here.
>
> It is unthinkable that the lower house of the legislature Saturday will withhold on final passage of the CVA bill the 100 votes needed to feed hungry, worthy workmen this winter.
>
> Texas is too big to permit any such cruelty to worthy citi-

zens. Let Congressman Buchanan tell the immediate benefits of passage of the bill by a two-thirds majority. A telegram he sent Cong. *(sic)* Fritz Engelhard Friday afternoon said:

"Bill as agreed in conference is satisfactory to Public Works Administration. If the Legislature will pass the bill as agreed by the conference committee by two-thirds vote, so it will go into immediate effect, I am confident that the loan can be consummated and about 1500 men put to work on the project in less than 60 days.

"I am further confident that I will be able to procure additional loans of from $8,000,000 to $10,000,000 from P.W.A. to construct other dams and power sites . . . and this will mean about 5000 additional people employed. If the bill does not take effect for 90 days, there is danger of P.W.A. funds being allotted to other projects in other states . . ."

The banner headline on the front page of the *Austin American* Sunday morning, November 11, 1934, proclaimed: "Texas Longhorns Smash Baylor Bears, 25 to 6." While the football game took top billing, the front page also gave due notice to House passage of the LCRA bill by a vote of only 75 to 40,[1] which meant it would become effective ninety days after the session adjourned. That made the effective date, and the official birth date of the LCRA, February 9, 1935.

One of the four decks on the one-column headline above that story declared: "Bitter Group of Die-Hards Cut Off Three Months of Winter Work." The article included these comments:

> Forty house members, led by a small group of bitter-end opponents, voted against the measure and thus destroyed three months winter employment for hundreds of people who could have been put to work. The law will become effective Feb. 9, under the vote of 75 to 39 which adopted a conference report.
>
> The CVA bill was saved from apparent defeat when Cong. Buchanan went to Washington and prevailed on Public Works Administration to accept a West Texas water priorities amendment previously ruled destructive of the loan. Opponents found another basis of continuing their fight when the conferees complied with the suggestion of Cong. Buchanan and struck out an amendment by Rep. Moore of Houston against the payment of

---

[1] According to General and Special Laws, Forty-third Legislature, Fourth Called Session, page 32, the official vote was 75 to 39.

any fees or compensation for financing the project or for sale of the property or securities . . .

And that was part of the news story on final passage. Adjoining it was another front-page editorial, which declared:

Texas will develop its own water power for the benefit of all electrical energy users. This was definitely decided Saturday when the House finally passed the Colorado Valley Authority bill 75 to 40 *(sic)*. While failure of the House members who opposed the bill at the behest of public utilities or for other reasons to give the bill 100 votes so that it could become immediately effective will be sorely regretted by all who need and deserve work on the project, the bill placed Texas definitely in competition with privately owned power producers . . .

Ablest of the utility lobbyists fought right down the home stretch to block the CVA. Their spokesmen to the very last were trying to make it appear that the program was designed solely to palm off on an unsuspecting public a worthless, uncompleted dam that is owned by a bankrupt Insull company. It was ever thus. Every serious effort of a public authority to provide any sort of competition for privately owned utilities is always met with a false cry that "it's wrong for the public to prosper, if, first any individual makes a dime."

Happily, all the senators but one and a substantial majority of the House saw the issue clearly.

Was Texas to reclaim water rights bartered away to Insull and use them for the benefit of the people? That was the basic issue. Most of the rest was a utility-inspired smoke screen . . .

Cong. Buchanan, whose tireless efforts turned the tide of victory from the utilities to the people, Sen. Walter Woodward, whose expert knowledge of Texas water laws and masterful diplomacy saved the bill from many pitfalls, as well as its authors and sponsors in both houses have earned the gratitude of those who know Texas can save its citizens millions in power bills and want to see it done.

Central Texas is undoubtedly entering a new era of cheap power. Its benefits will flow to all Texas in the future.

The original act simply created the Lower Colorado River Authority as a non-profit agency and gave it the power to utilize, store, develop, and sell the waters of the Colorado River and electricity generated by that water within a ten-county district. The LCRA was authorized to negotiate loans with federal agencies and to issue up to $10 million in revenue bonds. But it had no power to levy taxes or to grant remission of taxes. The bill included an appropriation of $5,000 for the agency's organizational expenses, with a specific requirement that this magnificent sum must be repaid out of the first revenue the agency received.

The Authority was authorized to issue revenue bonds to finance construction of facilities, but only up to the limits imposed by the legislative act (originally $10 million). None of its property could be mortgaged; only the revenues generated by the agency could be pledged for the retirement of those bonds.

And the LCRA was charged specifically with these responsibilities: (1) to control floods; (2) to conserve and store the waters of the river for useful purposes; (3) to develop, generate, and sell water power and electric energy; and (4) to protect the watershed land resources by aiding in the prevention of soil erosion.

As he reviewed the birth of the LCRA fifty years later, Sim Gideon noted:

> By allowing the LCRA to pledge the revenues that would be received from utilizing and selling the waters of the Colorado, the State received a chain of dams on the Colorado worth millions and millions of dollars.
>
> Actually, there probably were not any unappropriated waters in the Colorado River below where Buchanan Dam is now situated at the time the LCRA was created. Certainly there was not the amount that was assumed. The records of the Board of Water Engineers show there was not enough water going down the Colorado River to take care of the water rights of the rice irrigation companies.
>
> There were companies that had built their canals in the early 1900s and who had acquired water rights under the methods that existed then for acquiring them. But such rights were more than could be met by the waters of the Colorado River during certain years.
>
> It's one thing to have water rights — and another thing to have water.

The Forty-third Legislature's fourth called session passed the Brazos River Authority bill as well as the LCRA measure. It also established the basic method of discounts, still in use, for early payment of property taxes. But it refused to consider approval of additional relief bonds.

Exactly how much effect Jim Ferguson's arm-twisting contributed to the final passage of the LCRA bill will never be known. Mrs. Hughes still led the opposition, but the solid front presented in previous sessions by the Dallas delegation was shattered. The fact that the LCRA and centennial issues had been locked together was illustrated on the front page of the November 11, 1934, *Austin American*. Inserted in the article on final approval of the LCRA bill was a one-paragraph box headed: "Only Two Dallas House Members Back CVA." The article said simply:

> Of the Dallas delegation which sponsored Centennial legislation, two voted for adoption of the Colorado River report. These were Stinson and Savage. Rep. Reed and Rep. Hughes voted against adoption, and Rep. Parkhouse did not vote.

And the centennial bill? That same edition of the *American* reported, in summing up the special session, reported:

> All efforts to salvage legislation providing state financial assistance for the Centennial were futile. A bill appropriating approximately $3 million for this purpose passed the Senate but failed to receive consideration in the House. Another bill levying taxes of about $5 million for Centennial support was strangled by amendments in the House and never reached the Senate . . . The subject of appropriating money was a bitter bone of contention almost from start to finish.

Dallas finally had to settle for half a loaf during the Forty-fourth Legislature's regular session in 1934. Governor Allred insisted that the state was not financially able to provide assistance for such frivolous purposes as a Centennial celebration and threatened to veto any such spending measure. But he finally approved a $3 million bill which specified that $1,075,000 of that amount must be spent outside of Dallas County, including $250,000 for renovation of the Alamo and $250,000 for erecting a permanent memorial at the San Jacinto Battlefield.

By that time, the LCRA had gratefully accepted its $5,000 loan from the state government plus its authority to borrow from the federal government. It had, in fact, already begun parlaying that combination into a multimillion-dollar chain of dams.

7

"Fifty Hundred Thousand"

When most people hear that old story about the guy who said he had a million-dollar deal fall through at the last minute because "they wanted a hundred dollars *in cash*," they consider it a wild exaggeration. Judge Ferguson, recalling the $2.49 in cold cash that was necessary to close the LCRA's first $5 million bond transaction, figures that could have been an understatement.

Five million dollars seemed a long way off when the LCRA Act became effective on February 8, 1935. On February 9, the nine members of the original board of directors, appointed by the three state officials designated in the act, took the oath of office administered by Chief Justice C. M. Cureton of the Texas Supreme Court in the Court's library at the Capitol. Later, the new board posed for pictures in front of the Capitol with Governor Allred, who had succeeded Governor "Ma" Ferguson on January 15.

Allred's three appointees to the board were C. R. Pennington of Abilene, an insurance agent and director of the Retail Merchants Association; J. R. Key of Lampasas, a businessman; and Ralph W. Yarborough of Austin, an assistant attorney general who later became a district judge and then served in the U.S. Senate from 1957 until 1971.

Attorney General William McCraw appointed Judge Ferguson, then a young lawyer in private practice; S. Raymond Brooks of Austin, a member of the *Austin American-Statesman* staff; and A. J. Reinhard of Fort Worth, a U.S. labor conciliator.

Land Commissioner J. W. Walker named Roy Fry of Burnet, a druggist; Roy B. Inks, a merchant who was serving as mayor of Llano; and Fritz Engelhard, who had left the legislature after piloting the LCRA bill through the House.

Engelhard and Brooks had worked for many years on the Colorado project while Ferguson, Inks, and Fry had helped lobby it through the legislature.

Armed with a $5,000 loan and the authority to borrow, these nine men faced the awesome task of financing the purchase of the Buchanan Dam project with its lands and water rights, then getting construction resumed during the Great Depression. While the act specified that the LCRA's principal office would be in Austin, it failed to provide for office space or other facilities. The Travis County Commissioners Court came to the rescue and furnished office space in the county courthouse. And Attorney General McCraw loaned the board the services of his assistant, W. C. Davis.

At its first meeting on February 19, the board chose Fry as temporary chairman and general manager and Brooks as temporary secretary-treasurer. Then, turning immediately to questions already raised about the constitutionality of the LCRA Act, the board voted to ask for an attorney general's opinion. It also instructed Davis to prepare a formal application to the Public Works Administration for the promised loan and grant needed to complete the dam.

The board voted to employ as a stenographer Madalene Norris, daughter of John A. Norris, chairman of the Texas Board of Water Engineers. She became the LCRA's first employee, at a salary of $100 per month.

When the board reconvened the next morning, it was greeted by a telegram from Congressman Buchanan concerning the PWA questions about the agency's constitutionality. According to the board's minutes, the telegram said:

> Objections to the constitutionality of the Colorado River Authority just delivered to me by PWA attorney. It will be necessary to have a test suit filed and determined by the Texas Supreme Court in the most expeditious manner. Am tonight mailing you PWA's

memorandum of questions concerning constitutionality . . . A. J.
Wirtz is familiar with questions involved and I suggest you discuss subject with him after receipt of PWA objections in order
that the Board fully realize the extent and importance of the questions involved. Copies of all proceedings of the Board laying foundation for and filing of suit should be sent to me to procure approval of public works attorney on every step you take. This
includes copy of revenue bonds issued as basis of suit.

Ferguson attempted to have Wirtz employed immediately as
general counsel but his motion drew opposition from Brooks, who
wanted the appointment to be temporary, and from Yarborough,
who questioned Wirtz's eligibility to serve. Yarborough won adop-
tion of his motion to seek an attorney general's opinion on whether
any attorney representing gas and electric utilities, as Wirtz did,
would be disqualified under either the letter or spirit of the act from
serving as general counsel.

The subsequent opinion, written by Assistant Attorney General Davis, was issued on February 27 and concluded:

> The legislative body in placing the appointive power (of di-
> rectors) in three separate officials indicated an intense desire for
> an independent electorate. In refusing to hamper the directors in
> their selection of employees by restrictions, they indicate an
> equally sincere purpose that the Board should act with full re-
> sponsibility and with perfect freedom from political, selfish or
> private interest.
>
> It is the opinion of the writer that under all the facts attend-
> ant to the passage of this law, together with the law itself, that the
> Board may in its judgment employ whomsoever it may choose as
> its attorney. Respectfully, I would add that the full responsibility
> for the accomplishment of the purposes set forth in the Act lies
> with the Board.

The board immediately hired Wirtz as its "general attorney"
and instructed him to: (1) proceed with the filing of the suit on the
constitutionality of the Authority, (2) to proceed with an application for PWA funds, and (3) in cooperation with the officers of the
board, to proceed with negotiations for both loan and grant funds.

The selection of Wirtz was logical, since he was so familiar
with the problems facing the newly created agency. In addition to
drafting the first LCRA bill and redrafting succeeding versions, he
had represented the company which started construction of Ham-

ilton Dam and had been appointed its receiver in 1932. He not only was a specialist in the field of water law but was highly respected as an attorney, even by some of the officials in the PWA who were questioning the legality of the LCRA Act.

Finally, the board got around to electing permanent officers for the remainder of the calendar year. That took four ballots; three of them produced votes equally divided between Fry and Engelhard for chairman, with Brooks abstaining. On the fourth ballot, Fry was named chairman and Engelhard vice-chairman.

That first board meeting actually took place on eleven different days, with various recesses stretching it from February 19 through April 8. It became obvious rather quickly that the authorized limit of $10 million in bonds which the LCRA could issue was too low. It also became obvious that the bitter opposition to the LCRA which had accompanied its birth pains had evaporated. A bill raising the bond limit to $20 million was passed by the Senate, 23-1, on April 19, 1935, and by the House, 112-3, on April 23. Governor Allred signed the measure on April 24 and it became effective that day.

The increased limit and the lack of opposition encouraged the board members, but that first $4.5 million still seemed farther away than the 1,500 miles between Austin and the mother lode in Washington.

Before the LCRA was created, Secretary Ickes and Malott, as president of the Colorado River Company, had worked out an agreement identifying "immediate" and "ultimate" projects for the Colorado River. The immediate projects were the completion of the Hamilton Dam and reservoir, the construction of the hydroelectric plant and "other works for the purpose of flood amelioration, of the sale of water for irrigation, for hydroelectric power and for other programs." A loan of $4.5 million was to be made by the federal government for the "immediate" project should an appropriate public agency be created.

The "ultimate" projects would include

> a unified series of dams at and below Hamilton utilizing the water permits and properties held by the Company and C. G. Malott, impounding reservoirs, such hydro-electric works as may be found convenient and economical to utilize falling waters for the generation of electric energy for irrigation and other uses, transmission lines and other appurtenances; said unified system

is contemplated to be acquired, constructed and operated by a public agency.

The agreement also provided that "if a competent body of the State of Texas shall be created, it shall exercise available options and shall make application for a loan and grant to finance construction of said hydro-electric plant and/or other works necessary for the accomplishment of the 'ultimate' project, or any part thereof."

During the LCRA board's first lengthy session, it adopted a resolution approving in substance the terms of that agreement, including specifically those portions of the "Ultimate Project" with which it could reach agreement with the PWA. Malott was authorized as the board's agent in directing all matters in connection with the application and in negotiating the terms of loans and grants. He and Wirtz were told to proceed immediately with preparation of the necessary paperwork.

Wirtz and Ferguson were sent to Washington in March to inform Congressman Buchanan of the board's decisions and to seek his assistance in expediting PWA action on financing and the transfer of the Colorado River Company's properties to the LCRA. Ferguson recalled:

> We got on the train and went up there, and we managed to work out a lot of little quirks. The attorneys for the PWA had four legal points they wanted to clear up. They were questions involving the constitutionality of the LCRA Act. The only way we could get them cleared up was to file a friendly suit against the Attorney General in the Supreme Court, to mandamus him to approve some bonds.
>
> We were staying in the Washington Hotel. We rented a typewriter and took it up to Wirtz' room to prepare a bond transcript — an order authorizing the issuance of bonds, setting out the form and outlining the method of payment. He dictated and I typed. Then we took it over to the PWA and got them to approve it. They said they would accept that if the Supreme Court of Texas would clear it.
>
> A number of other things would have been a little draggy, but we managed to get them straightened out, too. We stayed pretty close to Buchanan and the office he had in the Capitol, as chairman of the Appropriations Committee. Lyndon Johnson and Wirtz were already close friends . . . Lyndon was very helpful even though he was just a congressional employee.

Johnson then was serving as secretary (administrative assistant) to Congressman Richard Kleberg of Corpus Christi and was working hard at laying the groundwork for his own political future. He was building a reputation as a Texan with contacts in high places.

Ferguson and Wirtz had trouble getting into see high-ranking government officials, and Johnson "went right to work on it and got us in to see everybody we needed to see," Ferguson remembered. "I know Wirtz and I both were very impressed with him. We left there feeling that he could get you in to any place." He continued:

> We were in Washington about a week. And while we were there, Mister Hunt, who was chief counsel for the PWA, and Carl Malott and Senator Wirtz all got to talking about the fact that there was a great deal of fall in the River and that there ought to be some other dams developed. There was something like a six hundred foot fall from the top of Buchanan to the bottom of the Austin Dam.
>
> Carl was also staying at the Washington Hotel — and we still had that typewriter up there. Senator Wirtz told me to sit down and write a memorandum about why we ought to make this a twenty million dollar project.
>
> Well, I sat there and used every artifice I could think of in trying to keep it down to one page. You had to keep your memo on one page or else they wouldn't read it. So I went over it and over it. I finally got it, presenting the idea that that much fall in the River could develop five times as much power as could be developed at Buchanan, and the cost would only be approximately four times as much.
>
> We took it over to Mister Hunt. He liked the idea. He had his secretary take that and make up a memo that was sent to Secretary Ickes. He approved it, too, so when we issued the first five million dollars, we already had approval for twenty million.

The total cost of that 1935 trip for Wirtz and Ferguson, according to LCRA board minutes of April 1, 1935, was $312.40.

> We brought the bond transcript back home and Senator prepared a very formal one then for the issuance of $100,000 worth of bonds. General McCraw got his bond man to turn it down on those four points they had given us, the ones they were concerned about. Then Senator Wirtz brought a mandamus action in the

Supreme Court to compel approval on grounds the objections were not valid.

> I remember when it was argued, [Ferguson added,] in the old Supreme Court Room in the Capitol, before all three judges and all six commissioners. Senator Wirtz made a very able argument and all of 'em were listening. General McCraw got up and said those four points were raised by people in a different place with a different idea of Texas law from his — but that he had relied on them to test them. He the same as told the Court he didn't believe that.
>
> After they came down off the bench, I remember Judge S. H. German saying to one group, "I'd like to see the *other* party to this lawsuit." The Supreme Court held with us on all four points and that cleared the way. It didn't take long after that before they sent us a contract from Washington. Then Senator Wirtz issued a real transcript for five million dollars in bonds.

Mother Nature even lent a hand in demonstrating the urgency of the situation. On May 19, 1935, one of the worst floods in history hit Austin. The river crested at 26.4 feet, the highest level recorded in Austin up to that time, as torrents of water came rushing down mainly from the Llano River watershed. The flood inflicted damages estimated at nearly $13 million on Austin and the Lower Colorado.

The Supreme Court decision was handed down three days later, on May 22, 1935. On May 25, the LCRA board held one of its most significant sessions. On that day, it adopted resolutions approving: (1) the loan and grant agreement with the federal government, together with the form of a revenue bond indenture; (2) the purchase contract between the Authority and the Colorado River Company; (3) a purchase contract with G. C. Malott for the twenty-eight parcels of land and the water permits he owned, and (4) a construction agreement with the U.S. Department of Interior's Bureau of Reclamation.

The board also was attempting to develop a permanent operating organization and had been informed by Ickes that it must select qualified persons acceptable to the PWA to serve as general manager, auditor, and treasurer. Lee C. Clark of Burnet was named treasurer and T. M. Markham of Austin as auditor. A committee of the board was appointed to receive applications for the post of general manager.

One thing upon which the LCRA and the PWA failed to agree

during negotiations for completing the half-finished dam was its name. A federal regulation prohibited dams from being named after living persons and Ickes steadfastly refused to recognize the popularly used "Buchanan" name for the structure. Federal correspondence kept referring to it as "Hamilton Dam" (Hamilton was still living, but federal policy permitted use of his name since the dam had been named for him while it was under private ownership) or as "Dam One." Minutes of the LCRA board indicate that the members continued to call it Buchanan Dam, except in such things as contracts with the federal government.

By mid-July, the LCRA's $5,000 loan from the state had been used up, but Governor Allred came to the rescue with a $5,000 deficiency appropriation.

Congressman Buchanan called Fry on July 20 and asked that members of the board, along with Wirtz, come to Washington immediately to meet with officials of the PWA. He explained that final details for closing the loan and grant agreement had to be worked out and a determination made on what actions should be taken by the board.

The speed with which things were moving is shown by the fact that an official meeting of the LCRA board was held in Room 900 of the Hamilton Hotel in Washington on July 22, 1935 — just two days after Buchanan's call. The official minutes show that board members Fry, Inks, Key, Brooks, Pennington, and Reinhard were present, along with Wirtz, Malott, and Secretary Ickes. After the necessary documents were approved, Ickes asked that a special meeting of the full board be held in Austin as soon as the members returned, to make doubly sure all their actions were legally approved.

The board reconvened immediately upon the return to Austin, on the afternoon of July 24, at the Wirtz firm's law offices in the Littlefield Building. After ratification of the agreements by the full board, the six members who had made that first trip to Washington headed back on July 25. They were to complete negotiations with the PWA and other federal agencies on procedures for getting construction under way and to get instructions for the required recordkeeping. Again the members were urged to proceed quickly with hiring a general manager to provide expert supervision of the construction program.

They also were persuaded by the PWA to contract with the

Fargo Engineering Company, which had prepared the original plans for Hamilton Dam, as consulting engineers. Fargo was to design changes for reduction of costs and "to make a complete development to impound water to the 1,020-foot elevation" level. The Bureau of Reclamation was designated to handle the completion of construction.

But first, of course, the LCRA had to get its hands on the money. That finally happened the first week in August, when the LCRA delivered the first $5 million in bonds to the PWA, which had agreed to buy a total of $10.5 million. The PWA also had pledged $5 million as an outright grant and the U.S. Reclamation Service had agreed to make another $5 million grant for flood control.

But in many ways that first $5 million seemed the hardest to collect. Ferguson, Clark as treasurer, and Gideon as Wirtz's law partner went to San Antonio to exchange the bonds for the money. Ferguson recalled:

> Wirtz took all the documents to Washington and got them approved. The PWA wired the money to the San Antonio branch of the Dallas Federal Reserve Bank, with the admonition that it was to be delivered in accordance with their instructions. They included a sample copy of the bond and said that all the bonds must be compared with that copy.
>
> A Mister M. Crump was manager of the Federal Reserve down there and he not only took things literally, but he was a stickler for detail. Mister Napier, president of the Alamo National Bank, also was there. They required that the trustee securing the bonds be a national bank with a certain amount of assets. None of the Austin banks could qualify so we had to get the Alamo National in San Antonio to act as trustee.
>
> I had delivered bonds for the City of Burnet to the Dallas Federal Reserve before and they would usually check a few of them without going over every one of them word for word. But Crump did it his way and had his people read every one of those $100,000 bonds against that sample copy. It took us about twenty-four hours over there.
>
> There were a bunch of papers Wirtz had been required to sign as our attorney — and he was supposed to sign every copy. Crump discovered one copy he had neglected to sign and he balked. Wirtz was on his way home from Washington and Gideon offered to sign it, as his law partner. But Crump said that wouldn't do.
>
> So we had to send a telegram to Wirtz asking him to send us

a wire authorizing Gideon to sign for him. The telegram was delivered to Wirtz on the train and he immediately wrote out the reply telegram we needed. But then he found out the train did not have another stop scheduled for some time. So he made it a collect telegram, and fixed it up with something to weight it down, then got the conductor to throw it off the moving train at the next telegraph station.

It arrived just a few hours after we'd sent him that wire and we all breathed a sigh of relief. One of the bank employees brought it in and gave it to Crump. But just as he started to read it, the employee said, "It's collect — two dollars and forty-nine cents."

Crump turned it face down and said, "Gentlemen, I'll have to have two dollars and forty-nine cents." Napier quickly said the Alamo National Bank would guarantee it. But that didn't mean anything to Crump. He said he had to have two dollars and forty-nine cents *in cash.*

We dug it up. I don't recall now who had it. I don't expect I did. Anyway, we gave him the two dollars and forty-nine cents and got our check for five million dollars.

Mister Crump handed me the check and it had "$5,000,000" on it in figures — but in the part where the amount was spelled out, it said: "Fifty Hundred Thousand Dollars." He apologized, explaining that his checkwriting machine couldn't write "million." And he asked me not to let anybody take a picture of that check. He was embarrassed.

I've thought about that many times. As much money as the federal government spends now — and back then they had a checkwriter that couldn't even write in the millions.

8

Calling a Bluff

A tragic aftermath of the rushed-up trips to Washington hit the LCRA Board of Directors on August 5, 1935, while it was in session preparing to close the deal with the Colorado River Company and its receivership. Roy B. Inks, who had contracted pneumonia on the second trip, died that day at his home in Llano. He was a native of Llano County, born less than a mile from the site of the dam which now bears his name.

The *Burnet Bulletin,* in a special edition on June 30, 1938, recalled "his untiring efforts to develop and harness the water power of the Colorado River." It noted that Inks was serving on the LCRA board at the time of his death and had worked "with characteristic energy and resourcefulness" for the completion of Buchanan Dam when the project "promised to be nothing but a white elephant."

On August 10, 1935, Land Commissioner Walker named Will Collins, who was editor of the *Llano News,* to fill the board vacancy caused by the death of Inks.

The day before, the LCRA officially bought the Hamilton Dam property during the "San Antonio Settlement." That transaction took place in San Antonio under the supervision of the U.S.

74

District Court that presided over the Colorado River Company receivership. Although the official appraisal of the properties was $3,798,058.97, the LCRA acquired them for a total of $2,630,959.11. Of that amount, the Colorado River Company received $1,275,347.22; Malott, $408,111.12; Fegles Construction Company, $814,498.10; Fargo Engineering Company, $49,149.94, and various loan companies and others, primarily for lands, $53,518.89. Legal expenses and miscellaneous small claims consumed $30,333.84.

The settlement represented a return of approximately twenty-two cents on the dollar for purchasers of the $3,714,500 in bonds which had been issued by the Central Texas Hydro-Electric Company. Morrison and his associates received a total of approximately $825,000.

Funds for the purchase came out of the $20 million commitment from the federal government, which included the purchase of $10.5 million in LCRA bonds by the PWA, a $4.5 million grant from the PWA, and the $5 million grant for flood control from the Reclamation Service.

Acquisition of the Hamilton Dam properties climaxed a remarkable series of accomplishments during the first six months of the LCRA's existence. Many of the achievements could be traced not only to the persistence and political finesse of Wirtz, and to the dedication of such board members as Ferguson, but to the calm leadership of Roy Fry in his dual role as board chairman and temporary general manager.

The Burnet pharmacist, a native of Council Creek in Burnet County, probably was as anxious as anyone to find a permanent general manager — and the PWA was getting awfully anxious. But the task of selecting the general manager produced the board's first real dispute with its benefactors in Washington.

The board made steady progress in filling other staff positions as it attempted to expedite the resumption of construction. On August 12, 1935, it hired as its land commissioner Ray E. Summerrow, whose efforts to help develop the Colorado dated back to his association with C. H. Alexander in Marble Falls. He was instructed by board resolution to take possession immediately of all the properties bought from the Colorado River Company and Malott, including all records, reports, applications, and the books of both parties.

The urgency of the instructions indicated a growing suspicion of Malott, who was trying to get the job as general manager. The fact that he was the brother-in-law of Morrison, whose dealings had provoked much of the opposition to the LCRA Act, did not work in his favor. But aside from that, most of the board members considered him a brusque, rather eccentric man and they especially resented the pressure he tried to apply on them through his Washington connections.

Ferguson recalls that he liked Malott "fairly well personally" but felt strongly he would not make a good general manager.

"He was a little bit 'Yankee-fied,' I guess," said Ferguson. "I don't know that he was politically ambitious, as some of the board members thought he was. I think his main objective was furthering Morrison as much as he could."

Hunt, the PWA lawyer, was the main force pushing Malott's bid for the job, according to Ferguson.

> Hunt got all of us to go to San Antonio in late July and he talked to each one of us individually — he wouldn't talk to us as a group. I think he intended — and this is kind of guesswork on my part — to really put the screws to us and make us hire Malott. If Secretary Ickes even knew about pressure to hire Malott, it would surprise me. He had to know a little but just what some of his staff had told him.
>
> I'm sure Malott thought I would be for him, but I knew he didn't get along with Wirtz and that would have been reason enough not to hire him. Nobody could have done that job if he didn't get along with Wirtz because Wirtz was too important to all of us. And you sure can't have two heads of an agency fighting each other.
>
> Malott had a lot of influence with Congressman Buchanan. Buchanan sent us a wire urging us to hire Malott, but I think that was just because he was so anxious to get the show on the road and he thought Malott was the man who could do it. Malott had been working closely with him on the Washington end of the thing.
>
> And Malott did a lot of favors for Buchanan. Mister Buchanan didn't have a car in Washington — I don't think he ever even drove a car — and Malott would go up there and rent a car, then drive him all over Washington. So I could see why Buchanan would be influenced more by him than by Hunt. But evidently, Mister Buchanan thought Secretary Ickes wanted Malott

and he didn't want to buck him on anything like that. He didn't want to hold up the project.

Ferguson believes that is why Buchanan sent the following telegram to the board on August 15, 1935:

> In order to get loan and grant I personally represented to the President, and he required it to be put in application to Allotment Board for Loan and Grant, that eight percent of the $20,000,000 must be spent by July 1, 1936. Good faith requires that such representation be kept. This will permit no delay in the progress of our project. General manager should be appointed at once and should be Malott or an engineer from Public Works upon recommendation by the Administrator of Public Works. Any other general manager not familiar with the project will cause unnecessary and dangerous delay.
>
> Importance of this project to Texas transcends all personal ambitions of applicants or desires of membership of your Board. We should sacrifice personal interests or feelings for good of project.
>
> The government relies solely on the revenues from this project for payment of the loan; therefore, it is vitally interested in seeing that every dollar is efficiently and economically spent and that the project be completed to a revenue basis within the twenty million dollar allotment. Public Works is determined to see that this is done and considers it vital to the solvency of the loan.
>
> Secretary Ickes is sending you a telegram. I strongly urge compliance therewith by the Board. Under existing contracts Authority now is incurring approximately $20,000 a month construction expense without receiving any value therefor because we have no general manager to direct operations.

The telegram from Ickes, as federal emergency administrator of public works, also dated August 15, said:

> I consider the appointment of Malott as general manager to be indispensable to the project's speedy accomplishment and financial success. Reasons for appointment have been stated to you by Mr. Hunt [PWA general counsel]. Therefore recommend you appoint him at salary of $12,000 per year. Your contracts with government require all appointments except auditor be upon general manager's recommendation; hence essential your Board make no appointments until you elect general manager including land commissioner, engineer and others. Please wire me decision of Board with regard to Malott.

Ferguson remains convinced that the wire signed by Ickes actually had been written by Hunt.

"Of course, it had to go out through Mister Ickes' office, but those higher-echelon people wrote things all the time and sent them around to his office to be dispatched," said Ferguson. "I just don't think he knew that much about this Malott thing, at the time."

After the telegrams from both Buchanan and Ickes were read, the board voted unanimously to reject officially the recommendation that Malott be hired. Ferguson described the action that followed.

> And then Senator Wirtz came up with a brilliant idea, like he usually did when something like that was happening. He recommended that Roy Fry, as chairman, send a wire to Mister Ickes saying we didn't feel Malott would make a good manager — but we were not trying to run this thing down here by ourselves and that we would hire anyone else he recommended, so long as he wasn't shown to be unfit for the position.

A resolution to that effect was adopted. It also said that Malott had been duly considered but the board felt that his appointment would cause friction and delay in the project. Not a single member could conscientiously vote for him, it declared, but everyone stood ready to accept an alternative recommendation from Ickes.

Telegrams summarizing the resolution were sent to Buchanan and Ickes. The one to Buchanan also said:

> We concur statement your telegram reference personal ambitions and sacrifice of personal interests and assure you we do not have any personal feelings or personal ambitions in matter, but would like to follow your suggestion if we could conscientiously bring ourselves to your view that Malott's selection would be good for project. Board being unanimously of opinion that his selection not advisable, has not exercised its rights under by-laws of making its own selection and submitting to Administrator for approval, but has called on Administrator for his recommendations which we think should eliminate question of personalities.

The board thus called Hunt's bluff, despite fears that the apparent stalemate might jeopardize the entire project.

Wirtz wrote a personal letter about it, emphasizing the political implications, to Lyndon Johnson. He knew that Johnson's influence was great not only with Kleberg and other congressmen but also with House Majority Leader Sam Rayburn of Bonham. In it, Wirtz said:

> I do not see anything left for the Board and myself to do decently except to resign rather than have Malott over us under the present circumstances. If we are forced to do this, you may know what a tremendous issue of interference in Texas affairs will be raised here. I think the appointing powers consisting of the Governor, Attorney General and Land Commissioner would accept the resignations and would enjoy making it one of the biggest political issues let loose in Texas for some time. I think our friends in Washington might desire to get a little friendly advice to the powers that be.

Wirtz also sent the following telegram to Charles E. Marsh, publisher of the *Austin American-Statesman,* who was visiting in Washington:

> Colorado Board unanimously against Malott for general manager on account of political intrigue and dictatorial attempt [to use] Authority for his own personal gains. Ickes sent H. T. Hunt [PWA general counsel] here on matter and he suggested Engineer Mitchell, whom Buchanan knows. Board agreeable. Malott returned to Washington and as a result Buchanan and Ickes now demand Malott be hired at $12,000. Ickes implies threat to whole project. Ask that you talk to Buchanan.

Fry, meanwhile, sent a telegram to Hunt which declared:

> Malott has attempted to dictate persons to fill positions which our counsel holds are not general manager's authority. This and other activities convince Board Malott is using politics to get absolute control of Authority for selfish and other ulterior purposes.

The barrage of telegrams to Washington brought a swift and surprisingly pleasant response, reinforcing the belief of Ferguson and others that Ickes actually had known little about Malott's campaign for the general manager's job. Said Ferguson:

> I think that's why he apparently turned around so fast. When we sent him a wire, he got it himself. There was one thing about him: he was an old curmudgeon, all right, but he was absolutely honest. When we put that to him, he couldn't very well do anything except what he did. And he really did us a great favor . . .

(Ferguson's "curmudgeon" reference was to one of the books Ickes wrote later, *Autobiography of a Curmudgeon*, published in 1943.)

Buchanan sent a telegram to the board on August 19, 1935, which said:

> I have conferred with Secretary Ickes who will later suggest to you first class engineer for manager Colorado River Authority. Little time required for Secretary to doubly assure himself man recommended is qualified in every way to handle this job.

Then, on August 24, Ickes recommended that the board hire Clarence McDonough as general manager, declaring that he had been chief of the PWA's engineering division since its establishment and had made an excellent record.

"I am willing to make the sacrifice of losing his services in the interest of your project," said Ickes. He added that it would be necessary to pay McDonough $15,000 a year, which in his opinion was "by no means excessive considering the responsibilities attached to the position."

Ickes said McDonough would go to Austin as soon as he received confirmation of the board's willingness to appoint him.

When Ferguson moved that McDonough be elected general manager for the period of construction at the $15,000 salary, Yarborough offered a substitute motion asking McDonough to come to Austin for an interview and conference about the position. The substitute failed and Ferguson's motion was passed with Yarborough and Pennington casting the only "no" votes.

Yarborough said he voted against the motion because he had had no opportunity to investigate McDonough and had no personal knowledge of him. "I may have no objection to him but I am voting 'no' for lack of opportunity to acquaint myself with him," he said.

Pennington said he shared Yarborough's feelings and added that another reason for his "no" vote was the salary. "I would not vote for any person at that large a salary," he declared.

McDonough arrived in Austin on August 31 with highly impressive credentials. A native of Gloversville, New York, he had attended Cornell University and the Massachusetts Institute of Technology, where he received an engineering degree in 1912. He went to work in 1913 for the Foundation Company of New York, a nationally known engineering and construction company. In 1918 he became engineer and district manager in charge of the Pittsburg District of the company. He served as chief engineer in charge of foreign operations from 1926 through 1930. When the Foundation Company hit hard times during the depth of the Depression, he became director of engineering for the Public Works Administration in 1933.

If he decided to take the position, McDonough said, he would be willing to remain with the LCRA until the construction phase was completed. He would also help select and train an operations manager before leaving. But he wanted a little more time to consider the matter and promised to let the board know his decision within a week.

A few days later, he agreed to take the job, effective September 18, 1935.

Ferguson and the other board members, after considering his qualifications and realizing the value of his connections with the PWA, were delighted. Ferguson summed up the choice:

> I don't think we could have employed anybody, anywhere, for construction that was the equal of McDonough. He had built dams all over the world and he knew what to do. And he knew the ins and outs of the Public Works Administration. He got us by a lot of things that probably would have delayed us, otherwise. He knew the people we needed — and he knew Mr. Alsop and got him to come down here and be superintendent of construction.

R. B. Alsop, who had been one of the top construction experts with the Foundation Company, eventually performed near-miracles in completing the construction of Buchanan Dam. But it took him a long time to get his turn at bat because the LCRA board had officially turned the Buchanan job over to the U.S. Department of Interior's Bureau of Reclamation in September of 1935, just as McDonough took over as general manager. And McDonough had to weather a deluge of legal and financial problems before the LCRA finally regained control of the Buchanan project ten months later.

A flood which roared down the Pedernales and Llano rivers, both of which flow into the Colorado below Buchanan Dam, had inflicted an estimated $16 million in damages on Austin and the Lower Colorado Valley in mid-June of 1935. That had prompted the LCRA board to go ahead with plans for a flood control dam three miles downriver from Buchanan, at the site which the Alexanders had chosen for a larger dam. The structure they had envisioned would have inundated some graphite mines in the area, which prompted Emery, Peck and Rockwood to decide instead on the Buchanan (Hamilton) site chosen originally by General Johnson.

The board later decided to name this second dam in honor of Roy Inks. But when McDonough came on board, no federal funds had been sighted for it and the Bureau of Reclamation had not even approved the location.

Meanwhile, that same flood of June 13–15 prompted Austin mayor Tom Miller and other civic leaders in the capital to begin serious discussions with LCRA officials on the possibility of rebuilding the Austin Dam and incorporating it into the agency's system.

Mayor Miller appointed a twenty-one-member Citizens Advisory Committee to work with the LCRA on that proposal. It was headed by E. H. Perry, a pioneer Austin developer, and one of the members was Herman Brown, the president of Brown & Root, Inc. Brown and his brother, George, a partner in the firm, were destined to play a major role in the growth of the LCRA — and vice versa. The fortunes of both the LCRA and Brown & Root were bolstered immeasurably by a man known as LBJ. And that, too, was a mutually beneficial arrangement which, ultimately, helped pave the road to the White House.

But the LCRA, during McDonough's early days with it, seemed much closer to the "poor house" than to the White House.

So did the members of its board of directors. They were authorized to receive only a $10 per diem "for each day spent in attending meetings of the Board," with no provision for travel expenses. The frequent board meetings, with some lasting as long as three days at a time, began to cause financial hardships for the members. But that matter was taken care of with a surprising lack of opposition during the first called session of the Forty-fourth Legislature. On October 17, 1935, both houses approved a bill providing $10 per diem *plus expenses* — so long as the board members were attending to authorized business and no member was paid per diem for more than 150 days during a year.

The same bill appropriated another $10,000 as an outright grant for the use of the Authority. Apparently, the vociferous opposition to the idea of the LCRA had evaporated.

When McDonough officially became general manager, the

LCRA had a bank balance of only $77,816 — none of which could be spent without special permission from the PWA. The agency had sold $7.7 million in revenue bonds to the PWA but had turned back $5 million of these funds to the Bureau of Reclamation, then paid the Fegles Construction Company and its creditors a total of $2,622,184.

One of McDonough's first official acts was to obtain from Ickes, his former boss, permission to spend the $77,816 for administrative purposes. Then he requested an advance grant of $750,000 from the PWA. That was received on November 20, providing the money necessary to proceed with the project.

Meanwhile, McDonough was spending much of his time badgering Bureau of Reclamation officials to get on with the job of completing the dam. He feared, among other things, that funds which had been allocated under the Emergency Relief Act of 1935 would be lost unless they were expended promptly on construction.

McDonough hardly had his feet wet when newspaper reports, during October of 1935, indicated that the $5 million allocated by the Bureau of Reclamation for flood control had been cut to $2 million.

Congressman Buchanan found that the problem was caused by a clause in the loan and grant agreement which said, "if borrower shall receive any funds directly or indirectly from the federal government or any of its agencies to aid in the construction of the project, to the extent that such funds are so received, the grants shall be reduced."

At that time, the LCRA had concluded arrangements with the PWA to use federal funds for the clearance of reservoir lands. But Buchanan obtained a ruling from the PWA that the clearance work was being done voluntarily, without any obligation involving the loan agreement, and therefore its cost would not be deducted from the grant.

A flap also had developed over the naming of an auditor. Markham's appointment on May 25, 1935, had never been approved by the PWA. In August, the PWA declared that Ralph W. Hunt would become the LCRA's auditor at a salary of $4,500 a year. On August 5, 1935, the board of directors accepted that decree but friction between Hunt and the directors, as well as the staff, developed quickly.

According to Ferguson, the board members made it clear to

Hunt within a short time that they disapproved of both his business and personal attributes. Hunt, apparently realizing that he was likely to be fired, responded by reporting to Washington that the LCRA board was attempting to circumvent PWA regulations and get more federal money than it was entitled to, according to Ferguson.

> The FBI became involved because Mister Ickes seemed to have a feeling that everybody was trying to "get to" the federal government. At least several of us on the Board of Directors had somebody following us for several weeks. I know I had somebody trailing me for a while.
>
> The Interior Department had some sort of problem in the Rio Grande Valley involving federal funds for irrigation ditches. I guess Mister Ickes thought all Texans were on the grab. Actually, we never tried to do anything that violated our agreement with the PWA. Now, we frequently discussed getting some provision of that agreement waived, and maybe that's what Mister Hunt told them about. But we never had any idea of doing anything without the approval of the PWA. We were moving pretty fast, at that point, as I recall, and there were quite a few times when the governmental red tape held us back.
>
> But it didn't take us long to discover that Mister Hunt was not the type of auditor we needed . . .

McDonough made the following statement on November 2, 1935, to Joseph E. Bishop, Jr., of the Federal Bureau of Investigation:

> I was in Austin, Texas, several days around Sept. 1, 1935, when the position of General Manager was offered to me. At that time I observed that the entire organization of the Authority was in turmoil and that neither the auditor nor anyone else was apparently doing anything to straighten matters out.
>
> Upon my return to Washington, I requested that C. J. Maxey send someone to the project to set up an accounting and bookkeeping system for the Authority. J. R. Coughlin and E. S. Fuerst, assistant supervising project auditors, were sent to the project and, in my opinion, have done a wonderful piece of work in establishing a feasible bookkeeping and accounting system. I have observed no improper conduct on their part since I have been engaged in my present capacity. At all times their conduct has been proper and has cast no discredit on PWA.

They have developed no intimate social contact with any member of the Board of Directors or officer of the Authority.

They have not, insofar as I know, attempted to belittle, ignore or insult Ralph W. Hunt, auditor to the Authority, at any time. Nor have they made any overtures to the Board of Directors in an effort to secure for themselves or any other person Hunt's position.

I have observed that the Board of Directors is making every effort to cooperate fully with the PWA and at no time have they voiced the attitude that once the Government purchased the obligations of the Authority that they would be able to manage the Authority's affairs without restrictions imposed by PWA rules and regulations.

No member of the Board has indicated to me that were Ralph Hunt replaced by some other individual the Authority would be able to escape the burdensome provisions of its agreement with PWA. The attitude of the Board of Directors, as I have observed it, is that while they are unable to pass upon Hunt's qualifications as an auditor, they have lost confidence in him and are prejudiced to him because of his conduct since he has been with the Project. Hunt has very injudiciously mixed in State of Texas politics, which has caused some dissatisfaction on the part of the Directors. He has assumed no responsibility or interest in the establishment of an accounting and bookkeeping system for the Authority.

I am personally of the opinion that Hunt is not a qualified accountant and is unfitted for his position . . . I do not believe Hunt has read or understands the provisions of the Trust Indenture. I am satisfied that he does not know the rules and regulations under which the authority is operating . . . He apparently does not understand PWA forms . . .

I have absolutely no confidence in Hunt's ability and work . . . I am definitely interested in having a change made. I believe it is in the best interest of the Public Works Administration and the Authority to secure a first class auditor.

Similar sentiments were expressed by Director Will Collins in another statement given that same day to FBI Agent Bishop. Collins said the board had lost confidence in Hunt because he had refused to cooperate with the members and added that his attitude had caused considerable friction. Hunt had given the false impression, Collins declared, that the board had tried to exercise a free hand in the management of LCRA activities without any regard to PWA policies. "Further," said Collins in his statement,

> it is not my position nor the position of any other Board member that with the replacement of Hunt by some other auditor acceptable to PWA there might be less stress laid on the requirements of PWA. In this connection, the attitude of the entire Board is to cooperate fully with PWA, the interest of PWA and the Authority being one and the same . . .
>
> I understand that prior to the employment of Hunt, Congressman Buchanan wired the Authority that Hunt was being sent to LCRA by PWA as auditor and that his salary was fixed at $4,500 per year. But that on his arrival, Hunt stated to the amazement of the Board that it was the understanding both of him and PWA that it would be $7,500.
>
> Further, he had made promises of permanent jobs to certain personnel after the Board had told him specifically not to do so. Further, in my opinion, he had been playing politics, or trying to do so, with the Governor's Office since his appointment. Also, when the Board wanted all equipment purchases to be under the General Manager, Hunt proceeded to purchase large amounts of furniture.

The Hunt problem was resolved suddenly and unexpectedly when he submitted his resignation, effective immediately, on January 3, 1936. That same day, the board appointed Coughlin to replace him — at a salary of $7,500 and with the authority to hire and fire employees of his department as he saw fit.

"Coughlin had been educated for the Catholic priesthood," said Ferguson, "and then he decided he didn't want to be a priest. So he went to law school and got a degree. After that, he became a certified public accountant. And he set up one of the best accounting systems that could have been set up for us."

Meanwhile, Ralph Yarborough resigned from the LCRA board in January 1936 to accept an appointment by Governor Allred to be judge of the 53rd District Court in Travis County. Allred named J. F. Hutchins, who was manager of a large rice farming and ranching enterprise in Wharton County, to replace him.

On January 2, 1936, the board officially named the dam to be constructed at what was known as the "Arnold Site" in honor of

Roy B. Inks for his "unselfish and unstinted service" in behalf of the LCRA.

In February, the board approved an amended Loan and Grant Agreement with the PWA which increased the amount of the PWA grant by approximately $1.1 million. That also made it possible for the LCRA to take over the construction of Buchanan Dam and let the Bureau of Reclamation assume the responsibility for building "Dam Three," the Marshall Ford project near Austin.

Finally, after its stormy birth, the LCRA appeared to be headed for at least a bit of smooth sailing. But then, the privately owned power companies went to court during the spring of 1936 with cases threatening the existence of many publicly owned, non-profit utility agencies — including the LCRA.

Suddenly, the battles which had been fought and won with Washington up to that time seemed minor. A war for survival had begun.

9

A "Hole" in the Dam

Early in 1936, the major utility companies operating in Texas filed a suit against Secretary of the Interior Ickes to stop payment of federal money to the LCRA and the Brazos River Conservation and Reclamation District for their construction projects. The plaintiffs included the Community Public Service Company, the Texas Power and Light Company, and the firms comprising the Electric Bond and Share Corporation (EBASCO). They obtained a temporary injunction from the Supreme Court of the District of Columbia (later renamed the District Court of the District of Columbia) prohibiting the federal government from disbursing funds to the two river authorities unless the power to be generated by their projects was to be sold to private enterprise.

The suit charged that a number of small towns, such as Liberty, Huntsville, Plano, Childress, and Gainesville, had entered into a conspiracy to buy electricity from the two river authorities. This was part of a nationwide movement contesting the concept of public financing for electric power generation facilities. A similar suit was filed by Duke Power and Light Company to halt public financing of a power dam in South Carolina, claiming that the use of federal funds for a power dam at Buzzard Falls on the Saluda River was illegal.

The plaintiffs contended that Title 2 of the National Industrial Recovery Act, which became widely known as the NRA and which authorized President Roosevelt to create the PWA, was unconstitutional. They claimed that the PWA loans also were unconstitutional because they made possible construction projects which could destroy the electricity business of private firms.

The temporary injunction was overturned so quickly that the LCRA never suffered any stoppage of funds.

"It just so happened," recalled Judge Ferguson, "that we didn't need any federal money while that injunction was in force."

But the threat lingered, spurring the Authority to act probably faster than it would have otherwise — not only on Buchanan and Inks dams but also on the Marshall Ford and Austin dam projects. And there was a rash of injunctions — five in all — issued during the long legal battle, but none of them actually hampered the LCRA's construction program. In one instance, according to Ferguson, that might have been only because the plaintiffs' lawyers "didn't know the territory." He called it the "hole-in-the-dam" agreement:

> Our folks agreed that the plaintiffs could have an injunction which would prevent us from completing one particular section of Buchanan Dam, thirty feet wide. The part we agreed not to build, though, according to the specified elevation and legal description, would have been on top of a mountain just north of the river. We agreed to leave "a hole in the dam" — on top of an adjoining mountain.

Fry, McDonough, and several LCRA board members gave depositions in connection with the suit, contending that the Colorado River projects were aimed primarily at flood control and conservation of water for irrigation, with the generation of electricity only a by-product.

Fry's deposition, dated May 27, 1936, noted that he had been interested in the development of the Colorado River "and particularly Buchanan [formerly known as Hamilton] Dam" for many years. There had never been any secret about the proponents' attempts to harness the river and to obtain federal funds for that purpose, he said. His deposition went on:

> Although it has never been the purpose of those sponsoring

> the project or the Board of Directors of the Lower Colorado River
> Authority to make this primarily a hydro-electric project and to
> so design, construct and operate the various dams as to give pri-
> mary importance to the generation and distribution of hydro-
> electric energy, it has been publicly stated that a quantity of elec-
> tric energy would be generated as an incident to the construction
> and operation of the dams as flood control projects; and in con-
> nection therewith many persons stated that this would result in
> competition by the United States Government with private inter-
> ests engaged in the electric industry. Representatives of some of
> the power companies, including some of the plaintiffs, opposed
> the enactment of the [LCRA] legislation on this ground, and the
> project was opposed in editorials appearing in some of the daily
> newspapers of the State on the ground that it would result in the
> confiscation of the properties of privately owned utilities through
> the sale of power to their customers by a publicly owned and op-
> erated body, at a low rate . . .

For at least two years, said Fry, the plaintiffs had known "or
should have been informed through the public discussions and the
press" that the federal funds to which they were objecting would be
used to complete Buchanan Dam and to construct other dams on
the Colorado "and for the construction of hydro-electric plants for
the generation of electric energy as an incident to the flood control
and irrigation features of the project."

But they failed, he declared, to lodge any objection with the
LCRA board or to take any other legal steps to prevent that construction

> until the filing of this bill of complaint, during which time the
> Federal Government and the Authority have expended large
> sums of money . . .
>
> The Authority does desire to proceed with the construction
> of the dams in order to provide flood control and prevent the
> enormous loss in human lives, destruction of livestock, agricul-
> tural products, and other property, caused by the flood waters of
> the Colorado River, which damage has averaged approximately
> $4,000,000 per year since the year 1900, and during the year 1935
> amounted to approximately $10,000,000 . . .

Any delay in construction, Fry added, would probably result
in additional flood damage and would impair the LCRA's ability
to pay interest on bonds held by the federal government. Also, he
said, "a large number of persons on relief will be deprived of a

source of employment and continue to be a matter of expense to the government."

McDonough's deposition echoed Fry's sentiments but also noted that he had reviewed the project while serving with the PWA and had turned down a loan initially only because it was then being sought by a private firm. Said McDonough:

> At no time during my service with the Public Works Administration or during my service as General Manager of the Authority have I heard anyone in authority with either of said organizations express a purpose of designing, constructing or operating the project in such manner as to provide for the generation of hydro-electric energy primarily and in such quantities as to destroy the market in Texas for electric energy or for the purpose of destroying the complainants or their properties or business.
>
> On the other hand, the persons representing the application of the Authority to the Public Works Administration have insisted that the project should be financed by the Government as a flood control project with aid to irrigation and navigation, and particularly pointed out that the prevention of flood damage of approximately five million dollars per year average, including damage to trunk line railroads and State and Federal highways carrying interstate commerce, would alone warrant the expenditure of Federal monies . . .
>
> It is my opinion that there is a sufficient available market for all hydro-electric energy which will be generated by the Authority without competing with the complainants. It is apparent from the bill of complaint that complainants operate in only a portion of the territory of the State of Texas, and that there is a large extent of territory in close proximity to the proposed project in which none of the complainants operate or have customers . . .

Ferguson recalls that McDonough, knowing exactly what the opposing lawyers were trying to prove, did an excellent job of clouding up their key points. In one instance, he said, the plaintiffs' attorneys were dealing with the question of whether or not the Colorado was a navigable stream. They made the mistake of asking McDonough about oyster beds near the mouth of the river.

"He talked for about an hour on oysters," said Ferguson, "until finally one of their lawyers turned to the other and asked, 'How do you turn this guy off?' "

Ferguson spent much of his time visiting officials of the various cities named in the complaint. He obtained statements from them to the effect that they had been planning to build their own utility systems and had not had any contact with the LCRA prior to the filing of the suit, thus combating the conspiracy charge.

The court suit dragged on for nearly two years without being tried. It finally was dismissed following a U.S. Supreme Court ruling on January 3, 1938, in the *Duke Power Company vs. Greenwood County, South Carolina* case. The Court, in a unanimous opinion written by Justice George Sutherland, upheld the constitutionality of the PWA and its right to allot a total of $2,852,000 to the dam at Buzzard Falls. The Court also declared that the private utility company had no immunity from lawful competition by the county.

That same day, the U.S. Supreme Court issued another ruling which would prove important to the LCRA several years later. In the case of *Alabama Power Company vs. Ickes,* the Court held that PWA loans to municipalities were legal and said the municipalities were free to construct their own distribution systems.

While the legal battle really did not slow down the construction projects, foot-dragging in Washington did — especially during the early stages.

Congress had authorized the completion of Buchanan Dam but had not provided the funds for it when, shortly after celebrating his sixty-ninth birthday on April 30, 1936, Congressman Buchanan made a visit to President Roosevelt. Buchanan later told Ferguson and Wirtz that the president was in an expansive mood and asked, "What can I do for you, Buck?"

"I want a birthday present," Buchanan told him.

"What do you want?" asked the president.

"My dam," said Buchanan.

"Oh, Buck!" Roosevelt is said to have replied. "Well, I guess we'll have to give it to you."

And with that, according to Buchanan's story, the president telephoned Ickes and told him to put the necessary funds in the budget.

On June 22, 1936, the LCRA board approved a contract with Morrison-Knudsen, Inc., of Boise, Idaho, for the construction of Inks Dam. And on the same day, the board authorized General Manager McDonough "to take full charge and to have full control, direction, and supervision over the performance of all construction work and the purchase of all materials necessary for the construction and completion of the Hamilton (Buchanan) and Inks Dams."

The patience of the board with the Bureau of Reclamation had reached the threadbare stage in connection with the Buchanan

Dam project. Fifty years later, Ferguson recalled well the frustration he and other members of the board felt:

> Eight months after the Bureau of Reclamation took over the project they hadn't done anything except sort of clean up the site and get the buildings back in shape — and we were in too big a hurry to go along with that.
>
> McDonough, being an engineer, kind of wanted to build that dam, anyway. He suggested that we see if we couldn't get it out of the hands of the Bureau of Reclamation and take it over — just hire everybody and build it ourselves. So we went back to Washington and, with Congressman Buchanan's help, we didn't have much trouble getting them to turn it back to us.
>
> While we were up there, Mister McDonough said a man named Bob Alsop, whom he had worked with before, was in Washington and he thought we ought to see if he would be interested in coming to work for us as superintendent of construction. McDonough had worked with him before on several projects and Alsop had just finished electrifying the Pennsylvania Railroad from Washington to New York. We hired him while we were up there.

The Bureau of Reclamation relinquished the Buchanan Dam construction job to the LCRA on July 1, 1936. Five days later, McDonough and Alsop had LCRA workers pouring concrete.

"Alsop organized the thing so well that there were times when he was pouring as much as two thousand yards of concrete a day," said Ferguson. "It was really an operation. And they finished that dam in just — actual working time — a little less than a year."

Among those fascinated by the skills of both McDonough and Alsop was R. A. Lucksinger, who was scheduled to begin working for the LCRA as a cost accountant at Buchanan Dam on July 1, 1936. His starting date was delayed ten days by the death of his wife's mother, but it did not take him long to make up for lost time, since he and others in supervisory jobs frequently worked seven days a week. They had three 8-hour shifts working five days a week as the construction workers raced the clock under the direction of McDonough and Alsop.

Lucksinger, who retired as LCRA controller and assistant general manager on July 31, 1974, is one of many who credit McDonough with breathing life into the fledgling LCRA. He declared during a 1985 interview:

> [McDonough] is the one who put the LCRA on its feet and

got it moving. He was the best engineer and the best administrator I ever knew. He was just a tremendous man. He had the brainpower and the knowledge to put the whole package together. He was the father of the organization.

He was good at dealing with the "feds" because he came out of that organization, as chief engineer under Ickes. He had the entree there. They knew when he came in with something it was going to be good. And we borrowed more than twenty-one million dollars from the PWA.

Lucksinger noted that McDonough was responsible for redesigning Buchanan and Inks dams and also had a lot to do with the Mansfield and Austin dams.

When we built Buchanan we didn't have construction equipment like we have today. We had an old Caterpillar tractor and we finally put a blade on it and made a bulldozer out of it — but the blade had to be operated by two hand-cranks. The equipment just hadn't been developed. That came along with World War Two. When we started in July of '36, that dam wasn't half-finished. It was probably between fifteen and twenty percent finished. There was not a yard of concrete that had been placed over on the Burnet side of the river. On the whole river section, down in the river proper, nothing had been done.

Everybody said it would take five or six years for the lake to fill. But we closed off the dam in May of 1937, less than a year after we started — and it is a tremendous project. Then we got a flood in July of '37 and it just about half-filled that lake overnight.

McDonough and Alsop developed a great construction organization. They were both highly respected and both were well-liked. Alsop was a real dynamo on a construction job. And he could be a fireball at times. But the men all got along fine with him.

Lashed by the threat of a federal funds cutoff, McDonough and Alsop maintained pressure that enabled them to finish the Buchanan Dam project in near record time.

In the meantime they also managed to expedite preliminary work on the dam twenty-one miles downstream that would become the key flood control link in the LCRA chain — Marshall Ford, later to be named Mansfield Dam — and on the reconstruction of the Austin Dam, which would become Tom Miller Dam.

A lengthy controversy over the site for the Marshall Ford proj-

ect, labeled simply as "Dam Three" at the time, finally was resolved on August 17, 1936, when the board chose what was then
called the "Hughes" site. The board said it was necessary to start
construction immediately to take advantage of available funds and
that land at the Hughes site could be bought or condemned without
any action by voters. On the other hand, noted the board, the acquisition of necessary land at the alternative "Maxwell" site,
downstream, would require an election to amend the City of Austin
charter, resulting in a delay of at least seventy-five days. In addition, building the dam at the Maxwell site would make it impracti-
cal to reconstruct the Austin Dam.

The original plans for the Marshall Ford project called for a
dam 180 feet high and 4,000 feet long. It so happened that the
U.S. Bureau of Reclamation was completing plans about that
time for a similar dam on the Columbia River, at Grand Coulee,
in the state of Washington. The similarities allowed the Bureau
to adapt much of the Grand Coulee design to Marshall Ford,
which speeded up completion of construction drawings and the
call for construction bids.

According to agreements made with the PWA and the Bureau
of Reclamation, the dam was to be built under the direction of the
Bureau (which also was to furnish the steel, concrete, and other
materials) after the LCRA acquired the land.

A tremendous amount of legal and clerical work was involved
in buying all the land needed for the dam site and reservoir. That
kept Wirtz and his law firm, and particularly Gideon, busy examining titles and preparing transfers of title. But fortunately, few
condemnation proceedings were necessary since the Colorado in
that area flowed through a fairly narrow canyon with rural, cedar-
choked lands alongside that was used primarily for pasture.

Finally, on December 3, 1936, the LCRA awarded a contract
for the construction of the Marshall Ford project to a fledgling firm
called Brown & Root, Inc., which had teamed up with the Mc-
Kenzie Construction Company of San Antonio to submit a low bid
of $5,781,235. At the same time, the secretary of the interior agreed
to order a flood control study of the Colorado River below Austin
to determine if the proposed height would be adequate.

The Austin Chamber of Commerce and the LCRA staged a
gala ground-breaking ceremony for the Marshall Ford dam on February 19, 1937. Both organizations were delighted when Secretary
Ickes accepted their invitation to participate and, on the same trip,

to see the progress being made on the Buchanan (which the PWA still called "Dam One") and Inks projects.

When Ickes arrived in San Antonio, he was met by Governor Allred, Lieutenant Governor Walter Woodul, State Senator Houghton Brownlee of Austin, and LCRA board members Fry and T. H. Davis of Austin, along with Wirtz and McDonough. The official party went first to the Buchanan and Inks projects, before heading for Marshall Ford. There, ceremony participants which had congregated included about 100 members of the Texas legislature.

According to the *Austin American* of February 20, 1937, Ickes announced that the Marshall Ford Dam would be built in two phases: the first to a height of 190 feet and the second to a height of 265 feet. That would raise the total cost, Ickes said, from $10 million to $20 million.

That same issue of the *Austin American* noted that: "Part of the crowd got ahead of the honor guest's car on the way from Buchanan dam to the Marshall Ford site, and the dust cloud their cars kicked up was something to do justice to the dust bowl country."

This fact was addressed by Governor Allred in his closing remarks at the ground-breaking ceremony:

> Out of deference to our distinguished visitor, I hope not a person will leave these grounds until he is safely on his way. We regret that on our way over here he got behind some of you good people and could not ride in the comfort to which his high station entitles him. Of course, the fact that your Senator and I are riding in the same car with Mister Ickes has nothing to do with this plea!

The dust was not thick enough, even in those days of cars without air conditioning, to choke off discussions with Ickes about two legal problems which had put the Marshall Ford project's financing in jeopardy. The U.S. Comptroller General's Office had refused to pay the initial $5 million in grant funds for flood control and irrigation to the Bureau of Reclamation under the Emergency Relief Appropriation Act. That office discovered that Congress had never specifically authorized the project, as required; instead, the funds had been provided under the PWA loan and grant agreements worked out with Ickes, through the assistance of Buchanan. And Congress had adjourned its 1936 session before this problem was uncovered.

Secondly, the comptroller general found that the lands upon which the Marshall Ford Dam was to be built did not belong to the federal government. The 1903 act creating the Bureau of Reclamation specifically prohibited the use of federal funds for construction on non-federal lands.

That problem was complicated by the fact that one provision of the LCRA Act declared that "nothing in this Act shall be construed as authorizing the sale, lease or other disposition of its property . . ."

Thus, under federal law the Bureau had to own the land upon which the dam was to be built. But under state law, the LCRA could not deed the land to the federal government or anyone else. Even Buchanan's influence as chairman of the House Appropriations Committee and his assurance that Congress would grant specific authorization for the project during its 1937 session were not enough to gain release of the funds — until his efforts were reinforced by pressure from President Roosevelt.

The comptroller general finally and reluctantly issued the first $5 million but warned Buchanan that the second $5 million for Marshall Ford would *not* be approved until both the authorization and land problems were resolved.

These complications came as a severe shock to Brown & Root, which had hoped to use the $10 million project as a ladder into big-time building. Suddenly, the Brown brothers were gambling their financial future on the actions of Washington politicians they did not even know. And Wirtz warned them that, should some congressman blow the whistle, they might not be legally entitled to payment for the work they were doing. It is little wonder that the experience renewed the Browns' interest in politics.

Robert A. Caro, in *The Path to Power*, the first volume of his study of Lyndon Johnson, declared:

> So much larger was the dam than any previous Brown & Root project that the company would have to purchase $1,500,000 worth of heavy construction equipment, including one particularly expensive item — a cableway consisting of two steel towers erected on either side of the river, with, hung between them, huge cables along which ran a trolley from which buckets filled with concrete mixed on the cliffs could be lowered to the men pouring it into the foundations — for which they might never again have a use. "We had to put in a million and a half dollars before we could get a penny back," George Brown recalls. Nor

could this investment be recouped from the initial $5,000,000 appropriation. Their estimated profit on that appropriation, without the cost of equipment figured in, was just under $1,000,000; on the first appropriation, in other words, they would be losing $500,000. It was on the second $5,000,000 appropriation that their profit would be made. On this second $5,000,000, Herman Brown had calculated, he and his brother would make a profit of almost $2,000,000 — enough to cover the $500,000 and leave them with an overall profit on the dam of $1,500,000, an amount double all the profit they had made in twenty previous years in the construction business . . .

But although the Browns were gambling, the power of Buck Buchanan hedged the bet. "Wirtz was telling us that Buchanan would take care of things, that he would arrange it (the authorization) as soon as Congress reconvened (in 1937), and of course we had no doubt of that," George Brown said. "We had already seen what he could do."

Not surprisingly, it was Wirtz who figured out a way to solve the seemingly insurmountable legal problems. His plan called for a parlay of Buchanan's tremendous power, President Roosevelt's blessing of the Marshall Ford project, and the belief that the Comptroller General's Office would be happy for any excuse to get off the hook. Thus, said Wirtz, they needed congressional passage of a bill not only specifically authorizing the project but also validating the contract with Brown & Root; if Congress approved a contract for building a dam, he noted, no one was likely to challenge the president and the Appropriations Committee chairman on a technicality of land ownership.

The idea sounded fine to Buchanan and he promised to carry it out. He went to work on it immediately when Congress convened in January of 1937 and it was scheduled for routine approval by Mansfield's Rivers and Harbors Committee in March.

But Buchanan, who had been hospitalized with a heart problem for three weeks during January of 1936, complained to friends after the convening of Congress in 1937 that he was more nervous and tired than usual. On Monday, February 22, he was at work when he suffered a heart attack and was rushed to the U.S. Naval Hospital at Bethesda, Maryland.

His wife and his son, James Paul Buchanan, Jr., were at his bedside when he died that night at 10:40 P.M.

10

The Biggest Catch

Wirtz, Engelhard, Ferguson, and Lee Clark drove to Brenham together on Friday, February 26, 1937, for Congressman Buchanan's funeral. They did a lot of reminiscing about the South Carolina native who had served as a Washington County prosecutor for seventeen years and as a member of the Texas House of Representatives for another six years before going to Congress in 1913. And they did a lot of speculating about who might win the special election which would be called to choose his successor.

"Wirtz was really worrying about that," Ferguson recalled. "He asked me if I would run and I said 'no.' In the first place, I was scared to death of politics, and I didn't think I could do any good. Then he started talking about Lyndon, who was in Austin serving as state director of the National Youth Administration, and saying what a good congressman he'd make."

Actually, Johnson had gone to Wirtz's office in Austin on Wednesday, the day after Buchanan's death, to seek his support in the race. Wirtz, whom Johnson had named chairman of the NYA State Advisory Board, had long been extremely fond of the young man he had seen perform so effectively in Washington but who was a political unknown in most of the Tenth Congressional District.

99

With the U.S. comptroller general breathing down the neck of the LCRA, however, Wirtz knew he could not let mere friendship interfere with the need to elect someone with enough key congressional contacts to help bail Marshall Ford Dam out of its legal difficulties.

Thus, Wirtz assessed in his own mind not only the potential candidates' possibilities for winning the special (high man wins; no runoff) election but also the influence each might have in Washington. Although the *Austin American*'s first article listing those who might run did not even mention Johnson, Wirtz ranked him the top possibility in terms of Washington influence.

Initially, Johnson was inclined to go along with the other potential candidates and stay on the sidelines if Buchanan's widow decided to run for the remainder of his term. But on February 28, the day before she was to announce her decision, Johnson boldly announced his candidacy. That reaped a ton of publicity for the twenty-eight-year-old Johnson and apparently influenced Mrs. Buchanan's decision, revealed later that day, not to run.

Among the seven other candidates who entered the race were C. N. Avery, a prominent Austin businessman who had been Buchanan's campaign manager, close friend, and chief liaison with the voters for twenty-four years; Houghton Brownlee, the district's popular state senator; Assistant Attorney General Merton Harris, a former district attorney who had long had his eye on Buchanan's seat; Sam Stone, longtime county judge of Williamson County; and Polk Shelton, a well-known conservative Austin lawyer who said he was anxious to go to Washington and fight President Roosevelt's plan to "pack" the U.S. Supreme Court.

Wirtz personally opposed FDR's proposal to gain control of the Supreme Court by enlarging it, but he did not let that interfere with the winning strategy he prescribed for Lyndon Johnson.

Wirtz advised his young protege to base his campaign on full support for Roosevelt and his New Deal program. Johnson did. He went all out on that theme, papering the district with campaign literature declaring that "A VOTE FOR JOHNSON IS A VOTE FOR ROOSEVELT AND PROGRESS" and simply, "FOR ROOSEVELT AND PROGRESS."

Johnson contended that the election was the first test of voter sentiment on FDR's "Supreme Court Reform" plan. Wirtz's theory that this strategy might attract the support of Governor Allred, an ardent New Dealer, and also that of Charles Marsh, the *Austin*

American publisher, proved correct. Allred, ostensibly neutral, persuaded his own campaign manager, Claud Wild, to take a job handling Johnson's campaign; the governor also got his secretary of state, Edward A. Clark, who would become one of Johnson's closest associates, to raise money for LBJ.

Clark, during a 1985 interview, estimated that Johnson spent between $75,000 and $100,000 during that first race, for which no reliable records are available. Whatever the exact amount, it was enough. Johnson won with 8,280 votes. Harris finished second with 5,111, while Shelton had 4,420, Stone 4,048, Avery 3,951, and Brownlee 3,019.

Johnson scored his victory on April 10, 1937, but for Wirtz and other proponents of Marshall Ford Dam, the best was yet to come. One month later, on May 11, 1937, President Roosevelt finished a fishing trip in the Gulf of Mexico. Governor Allred had made arrangements for Johnson to meet the president and pose for pictures with him on the dock at Galveston, then ride to College Station with him on a private railroad car. President Roosevelt became so fascinated with the young congressman-elect that he asked him to ride another 200 miles, to Fort Worth, with him. By the time the trip ended, a young man who had not yet been sworn in as a congressman had become a protege of perhaps the most powerful president in the history of the United States.

FDR had gone fishing — but it was LBJ who made the biggest catch of all, without even getting on the boat. Wirtz himself could not have written a better script for LCRA purposes. Seeking influential contacts in Washington, he obviously had hit the jackpot by backing his dark-horse candidate.

Johnson arrived in Washington on May 13, 1937. He immediately went to the office of his old friend, newly elected Majority Leader Sam Rayburn; he flattered Rayburn by kissing his bald head and asking him to stand with him, as his sponsor, during the oath-taking ceremony.

Eleven days later, Mansfield's Rivers and Harbors Committee approved the bill Wirtz had written specifically authorizing the construction of Marshall Ford Dam and declaring that "all contracts and agreements which have been executed in connection therewith are hereby validated and ratified."

It took another two months to gain final passage of that important act. Johnson asked Wirtz and members of the LCRA board to come to Washington and help him with the final push. The LCRA delegation arrived on July 20 and two days later was invited to the White House, where James Roosevelt, the president's son, handed over papers approving the additional $5 million appropriation for the Marshall Ford project.

Johnson thus jerked the fat out of the fire not only for the LCRA but also for Brown & Root. Full financing finally had been achieved for the 190-foot-high dam at Marshall Ford. LCRA board members, along with Wirtz and McDonough, expressed their deep appreciation to Johnson; and, while they were there, they also called his attention to the additional flood control potential involved in increasing the height to 270 feet.

After returning to Austin, the board members adopted resolutions expressing appreciation to Congressmen Johnson and Mansfield and to President Roosevelt and Harry Hopkins, the works relief administrator. Hopkins earlier had voiced concern about the eligibility of the project for relief funds since it was being built by a private contractor with his own skilled employees, rather than by a governmental agency with workmen taken off the relief rolls. Without mentioning that particular problem, one of the resolutions praised Johnson's "diligence, energy and ability" in securing the second increment in federal funds.

Ferguson was especially grateful for Johnson's efforts. He later said:

> We had to have somebody up there. We were going great guns with Buchanan. He was so powerful, as chairman of the Appropriations Committee — it was amazing what those governmental departments could do when Buchanan asked them. Nobody else could get anything. But when Buchanan asked them, they'd fall all over themselves. Everything was so perfect with him — and then, when he was gone, the friends we had were not in a position to help.
>
> Lyndon had to take over under difficult circumstances. He had a hard row to hoe for a while and if it hadn't been for Roosevelt, he couldn't have done it. But Roosevelt was his friend and when he'd get in trouble, he'd go to the President . . .

Meanwhile, the LCRA faced another court challenge, this one from the largest irrigation company in the lower Colorado Valley — the Gulf Coast Water Company. When the LCRA refused to sell water for rice irrigation on terms acceptable to that firm, the company obtained a court order on July 28, 1937, requiring the Authority to release at least 1,500 cubic feet per second of water from the Buchanan reservoir for two weeks.

Wirtz appealed the order to the First Court of Civil Appeals and got it reversed on August 3, 1937. Justice Graves wrote the Court's opinion, declaring that the irrigation company was entitled only to protection against interference with the normal flow of the river and there was no evidence that this right was being denied.

The controversy finally resulted in a far-reaching agreement on rice irrigation. Board Member J. F. Hutchins of Pierce (Wharton County) invited representatives of Gulf Coast and the two other major irrigating firms, Lakeside Irrigating Company of Eagle Lake and Garwood Irrigating Company of Garwood, to meet with his committee and work out a settlement. When they did settle, the LCRA agreed to waive payment each year for enough water to irrigate the annual average number of acres actually irrigated by each company from 1930 through 1936. This, they decided, was the acreage which could be irrigated from the normal flow of the river. The 1930–36 period was chosen because it represented the period immediately prior to the closing of Buchanan and Inks dams.

In approving the agreement, all three companies recognized the rights of the LCRA to control, store, use, and sell the waters of the Colorado River. They stipulated that except for the water needed to irrigate the number of acres specified, they had no other rights to Colorado River water.

While LCRA officials haggled with the irrigation companies, they also continued to negotiate with the City of Austin on the reconstruction of the Austin Dam — and tried to expedite the completion of the Buchanan and Inks projects. When they learned that Secretary Ickes was planning another trip to Texas for mid-October, they decided to stage the formal dedication of both Buchanan and Inks dams on October 16, 1937, at the Buchanan site.

Mrs. Buchanan was too ill to attend the ceremony, but among the honored guests, in a crowd estimated at approximately 500, were Mrs. Roy Inks and her son, Jim Moss Inks, and her daughter, Mildred Inks. Wirtz served as master of ceremonies and Lieutenant Governor Walter Woodul, Austin Mayor Tom Miller, Senator Brownlee, and about forty members of the Texas legislature were among the dignitaries who helped welcome Ickes back to Texas.

Congressman Maury Maverick of San Antonio managed to get in a plug for Johnson, telling the audience that he had been happy to help Buchanan with the project "and also his very able and fine successor, Congressman Lyndon Johnson; I can't make a political speech here but I hope you will re-elect him."

Johnson's introduction of Ickes and the secretary's twenty-three minute speech, in which he castigated the privately owned utility companies which opposed public power projects, were broadcast on a statewide hookup. Said Johnson:

> These two projects are a monument to a fearless, courageous and positive leader, to whom the United States can never pay its full debt except in unalloyed gratitude and remembrance — a monument to Harold L. Ickes, Secretary of the Interior, Public Works Administrator, a man who lives on work, who turns dreams into tangibilities, and whose daring drive and force make national programs spring up in a thousand places at the same time.

Ickes expressed regret that the late Congressman Buchanan could not be present for the celebration and congratulated the people on choosing Johnson as his successor. His speech continued:

> This fine project, besides adding to our national wealth, is tangible evidence that you here in Texas, who have supported it from the beginning, subscribe to a new idea which in our generation has come to dominate the thoughts of forward-looking people about government. This idea is that the resources of the nation can be made to produce a far higher standard of living for the masses if only government is wise and resourceful in putting them to use for the common good.
>
> President Roosevelt has consistently advocated this idea and, during his recent trip to the Northwest, he gave vigorous and eloquent expression to it. The prudent development and wise use of our natural resources is an ideal always kept before them by the Bureau of Reclamation and the Public Works Administration in their planning.
>
> There have been attempts to discredit the conservation program of the present administration. Certain interests have attempted to put a stop to the great work that is being done to improve navigation, prevent floods and irrigate land that otherwise would remain unused.
>
> One reason for these attacks is that the government insists on using falling water to generate electricity. Another is that here,

this Authority — and the federal government elsewhere — have refused to discriminate against municipal systems in order that privately-owned power companies might make privileged profits.

When we look at a specific project like yours, we understand the situation better. Here, Texas citizens, seeking to control this river and make it a servant of the people, persuaded the legislature to establish the Lower Colorado River Authority. Unable to obtain financing from private sources, they availed themselves of the opportunity of obtaining federal funds. The government was willing to advance the money to furnish employment, to take part in the construction of a great conservation project which at the same time would be a permanently useful public work.

Why have the public utility lawyers fought so hard to stop such an enterprise as this? At the inception of this project, the private power companies believed that they would be able to purchase all of the electric energy generated, and in turn resell it to the people — thereby entrenching their existing monopoly.

The expenditure of public funds from the federal treasury for such a purpose was regarded by the power companies as neither improper nor unconstitutional. They welcomed such action by the federal government, provided only that they were to be allowed to fry the fat out of the profits from the sale of the power . . .

The basis of the opposition for some of them is not that they cannot purchase power, but that the municipalities can also purchase it. In other words, they want a monopoly. But the people want water and flood protection and, with power they can pay for them. They also want cheap power. If they cannot get it from private utilities, they will get it for themselves, as they have repeatedly demonstrated. And the Supreme Court itself has many times upheld their right to do so.

Multiple-purpose development of our rivers we cannot expect from the power companies interested primarily in the maximum profits they can squeeze out of power. In most instances, that must come from the federal government or from such an authority as you have set up here, whose primary interest is not in profits but in the full and diversified use of the river for the benefit of all the people.

Such a development as this must come from the cooperative efforts of the people themselves. Other ages are remembered by their pyramids over dead Pharaohs, or triumphal arches to conquering tyrants. I believe the distinguishing monuments of this era will be the great conservation projects, such as Boulder, Bonneville, Norris, Grand Coulee, Buchanan and Inks Dams, which the people are building in their determination to keep and use the great resources which nature has given us.

So long as this administration continues, its conservation program will be designed to benefit all the people and not to enrich the privileged few. If such a program can be carried out, it will go to prove that this democracy, if intelligently used, will work.

11

Power to
the People

Sim Gideon recalls going out into rural areas with Lyndon John-son during the future president's first campaign for Congress and watching him make speeches in small-town churches — by lantern light.

"He'd promise 'em he was going to get 'em electricity," said Gideon. "I thought, 'my Lord! He'd promise 'em anything to get a vote.' In those days, you'd support anybody for Congress if you believed he was going to get electricity for you. But I just didn't see how you were going to get electricity out to those ranch areas."

Gideon, having grown up in a rural area, near Coleman, always considered electricity as an impossibility in his area because it was too expensive to build the lines. "I couldn't conceive of building lines out there," he said, "where you have only four or five people to the mile, and getting electricity to them. But he did."

Lyndon Johnson knew first-hand of the awesome hardships experienced by farm families without electricity. According to Caro's *The Path to Power*, Johnson's most bitter disputes with his father, Sam Ealy Johnson, Jr., resulted from his refusal to help pump water from a well and haul it into the house for his mother. Without electricity, woodstoves provided the only source of heat; Caro noted that "Johnson's refusal to chop wood for his mother was another source of the tension between him and Sam."

Life without electricity is difficult for most people today to imagine. But even as recently as the late 1930s, many farm families — especially in the Texas Hill Country — had to draw all their water up by hand from a well, sometimes as much as 100 feet, and haul it into their houses. Many of them had never heard of indoor plumbing. Farm wives slaved at least one day a week over outdoor washpots and scrub-boards which left them wrinkled and bent long before their time. Most had never even heard of the radio, and the dim, flickering light of kerosene lanterns was not conducive to reading what little they had to read.

Still, when they first were given the opportunity to sign up for electricity, many balked. With the Depression still holding axes over their heads, they were afraid to sign anything that obligated them to pay money; forfeiture on even a small amount, they feared, might cost them their land. They simply were not willing to take that risk for what they thought, initially, would merely be a small light globe to replace a kerosene lantern.

Their reluctance helped promote the classic confrontation between public and private power suppliers. But many of the Hill Country farmers stumbling around in 4:00 A.M. darkness trying to milk their cows knew nothing better until Johnson described a promised land. Their wives did not realize how electricity could solve so many of their cooking and washing problems and provide refrigeration that would keep them from having to start every meal from scratch.

Johnson grew a bit impatient when the LCRA seemed unable to keep pace with his promises. In a letter dated March 16, 1938, he wrote to McDonough:

> You and I both realize there are plenty of rural communities throughout the whole area, where smokey lanterns are the chief means of lighting and elbow-grease is still the main motive power, although this is the Twentieth Century and not the Middle Ages. We know there are many towns in the Tenth District either entirely without electric light and power or struggling with inadequate, expensive, wasteful and cumbersome plants of their own, which supply light and power of most unsatisfactory kind at rates nothing less than blushful. There is no reason why either of

these conditions should longer exist in the whole 40,000 square mile area the Colorado River is getting ready to serve.

There is no program of more interest to me personally and officially than that upon which you and your associates have been working the last few years, and I want you to know that we must see to it that it is made entirely effective. I shall be happy to assist in spreading information on the organization of rural electrification cooperatives and corporations and to give these projects my full attention when they reach the Washington office for approval.

At McDonough's suggestion, the LCRA had just had a survey made by the Fort Worth engineering firm of Hawley, Freese and Nichols to determine the potential market for the 300 million kilowatt hours of electricity to be generated annually by the Buchanan and Inks plants. Since the Authority had to be self-sustaining, the sale of this power was vital to its survival.

The report recommended that the LCRA seek commitments for its electricity from municipally owned power plants within 200 miles of Austin, from rural cooperatives, from industries which did not have central station service, and then from cities and industries which had central system service.

Neither the LCRA board nor Congressman Johnson was pleased with the recommendations. They favored the philosophy espoused by President Roosevelt, feeling that the power should be sold to the people of the region at the lowest possible cost. While Buchanan had subscribed to the same idea, he was not as militant in pushing it as was Johnson. The new congressman urged the LCRA to move quickly in selling to municipalities and the rural groups which were beginning to organize rural electric coops — and to either buy out the systems of privately-owned utilities or establish systems to compete with them.

Johnson, as well as the LCRA, was feeling pressure from the parents of the New Deal to expedite the development of markets and begin generating the revenue necessary to repay the federal loans, and thus provide proof that the public power schemes were practical.

On March 21, 1938, the LCRA directors adopted the keystone of their power policy, a resolution which said:

Be it resolved by the Board of Directors of the Lower Colorado River Authority that it is the policy of the Authority to sell all of its power, if possible, to municipalities and other public agencies for distribution to the public, in order that the electric customers of this State may enjoy the utmost benefit of cheap power made available by this enterprise.

The resolution went on to declare it imperative that both the Authority and public agencies (including municipalities and cooperatives) follow "a vigorous policy in order that the Authority not be forced to the extremity of selling power to private agencies who may not pass the benefit on to the customer."

The board advised the PWA of the policy it had adopted and asked that federal funds be made available for the transmission lines and other facilities necessary to deliver cheap power to the potential customers. The official request noted "the grave danger of the Authority being compelled to sell a large part of its power to private agencies in order to meet its obligations on its bonds" unless the funds were provided for building lines.

About the same time, a conflict developed with the Brazos River Authority (BRA), which had been established in 1929. The BRA was seeking federal funds to help build the proposed Possum Kingdom Dam and trying to find a market for the hydroelectric power it planned to generate there. The BRA directors took a dim view of the LCRA's offer to sell electricity to municipalities and coops in both the Colorado and Brazos basins.

While the directors seethed quietly in their reluctance to ruffle the waters, the Temple Chamber of Commerce sent out telegrams to cities in the Brazos River basin urging them *not* to buy LCRA power. When Wirtz learned of that, he wrote a strong letter to John A. Norris, the BRA's general manager, saying the two agencies should not fight in public while the private power companies "enjoyed themselves from the sidelines."

Wirtz said the LCRA had power ready for immediate sale, but it would be some time before the BRA power was available. He also said it should be obvious that there was a market in Texas for all the power that could be generated by both agencies.

Wirtz concluded with a suggestion that representatives of the two agencies get together and work out a program of mutual benefit.

As Kenneth E. Hendrickson, Jr., noted in *The Waters of the Brazos, A History of the Brazos River Authority 1929–1979:*

Representatives of the two agencies met in Temple on May 17, 1938, and agreed to formulate a jointly acceptable policy for the disposition of hydro-electric power. During the next few months their negotiations went on while both attempted to reassure the public and the federal government that they were not at each other's throats. This process was much more difficult than it should have been because the press was not particularly helpful, but at length, in December, 1938, an agreement was reached whereby the [Brazos] District would sell power to the [Colorado] Authority for disposition to cities and other public agencies, and whereby both groups solemnly promised not to sell power to private utility companies. It was amicable, and would have been a workable settlement, except for the fact that it was based upon the assumption that the District would obtain additional federal support from the Public Works Administration. As it turned out, such funding never came and the cooperative agreement between the District and the LCRA died.

Congressman Johnson, meanwhile, felt that such minor problems as squabbles with another river authority should not be allowed to interfere with the task of getting electricity into the rural areas of the Hill Country. On April 12, 1938, he met in Washington with the LCRA directors and gave them what amounted to a lecture on salesmanship.

Johnson told them the Rural Electrification Act, which had been passed mainly due to the persistence of Congressman Rayburn, could be used to help finance transmission lines into sparsely settled ranch country where such lines could never be made to pay for private utility firms. He was unhappy that the board seemed so far away from even applying for a loan to finance such lines. He recommended that the LCRA make a market survey to determine potential customers within a 250-mile radius of each dam. Every town should be given a detailed explanation, he said, of how it could obtain LCRA power and how much money it would save them.

"I am convinced that the present high electric rates in Texas are inexcusable and beyond reason," Johnson declared. The LCRA was in a position to drive those rates down, he added, but could not do so if it wound up selling most of its power to privately owned utilities. He continued:

> Rural electrification offers the Authority the opportunity to put into effect the principles and aims behind the total Adminis-

tration program in which it has been given a chance to exist in the first place — a control of floods and the turning of power into channels for a better life for farmers, ranchers and residents of small cities.

The directors appeared to take the pep talk to heart and vowed to get things rolling as soon as they returned to Austin. Johnson resumed his efforts to gain more federal assistance for the LCRA and came through with $2.35 million for the completion of current projects, $5 million for the construction of transmission lines, and $2.35 million to be applied toward retirement of outstanding bonds.

But when he detected no signs of fast action by the LCRA directors, Johnson wrote a letter dated May 12, 1938, to Chairman Engelhard:

> When members of the Board of LCRA were in Washington, I was assured that applications for additional funds for the Marshall Ford Dam and for the construction of the transmission lines could be worked up within two weeks. That was on April 12; this is May 12 and applications have not yet been received. Furthermore, I have no idea when they can be expected.
>
> At the same time, I was assured that I would receive regular weekly reports on contacts made with the cities concerning the sale of power, the prospects for the sale of power, and other steps in your program to market the energy we must sell to meet our obligations. Four weeks have passed and I have not had one scratch of the pen on the subject.
>
> Because of the assurances I received, I went ahead and did a great deal of the necessary preliminary work with the Public Works Administration. I laid plans to get things in the mill the minute you folks are ready to go.
>
> I regret to have to say once more that because you folks have not done the jobs agreed, everything is hanging in the air. The only prospect that I can see is the last minute rush in Congress when everything gets jammed up and will run the risk of going completely haywire.
>
> I have received telegrams and letters every week from towns and cities which I presume you have contacted on the question of power. I do not know, of course, how to answer these inquiries since I have not heard anything from you.
>
> . . . I feel that I must say to all the members of your Board that if they expect further assistance from me in working out their problems, I shall have to have more cooperation than I have received the last four weeks.

I feel that if the trouble lies in insufficient office and field help, additional help should be employed immediately so that the gap can be filled without further delay.

I have devoted a great deal of time and effort to the work of the Lower Colorado River Authority. I do not feel that I need to go into my interest in it or my expectation of the good that it can do for the District, for all of Central Texas; but I am constrained to say that unless it can keep its agreements with me, I shall have to let it resort to its own resources. I wish you would let me know right away what I can expect.

Needless to say, the apologies were quick and profuse. And so was the action by the LCRA Board of Directors. The board not only began efforts to determine its market potential and organize rural electric coops but also to pursue its original goal of an "Ultimate Project" involving six dams. It instructed McDonough to proceed with the initial planning of dams and power plants at two sites where the Authority had acquired water permits from the Syndicate Power Company. These would eventually become the Alvin Wirtz and Max Starcke dam sites.

But as the start of construction on the Austin Dam neared, the LCRA's first labor union problem appeared. It involved the International Brotherhood of Electrical Workers (IBEW). In late May, McDonough and D. A. J. Sullivan, of the PWA's Labor Relations Division, completed an agreement with the IBEW's Austin Local to facilitate construction of the power plant and its equipment. The agreement, which was approved in June by the board of directors, was to apply *only* to construction work and would "in no wise apply to the operation of the power plants." It was to become effective June 11, 1938, for one year, and continue thereafter on a year-to-year basis until cancelled by either party.

It specifically provided that there would be no work stoppages, either by strikes or lock-outs. The union agreed to furnish competent workmen, and the LCRA agreed to hire only union men when possible.

The LCRA also reached an agreement with the City of Austin for reconstruction of the Austin Dam, which eventually would be named "Tom Miller Dam" in honor of the city's longtime, highly respected mayor. The Authority's application for federal grants and loans of approximately $14 million for that project was approved by the PWA in a contract dated July 1, 1938.

Full-scale reconstruction started on July 5, with the long-jinxed dam to be built 100 feet high and 1,590 feet long. The spillway crest was to be 492.8 feet above mean sea level and a concrete apron would extend several hundred feet downstream in order to prevent the type of undermining action which had wrecked the dam in 1900. The project would provide a total of 1,400 jobs during the two-year reconstruction period. But disaster had to get in one more lick before the project was completed.

During the last week of July, a tropical storm battered the Edwards Plateau region of the Colorado River watershed and produced cloudbursts which wreaked havoc at Brady, San Saba, and Bend before the floodwaters joined the main river above Buchanan Dam. The storm also dropped near-record amounts of rainfall on the watersheds of the Llano and Pedernales rivers, which flow into the main river below Buchanan.

At Austin, the Colorado River produced a flood with the highest volume of water ever recorded there.

The level of Lake Buchanan was 1,012.2 feet at the time of the first recorded inflow into it on July 21. This was 7.8 feet below the normal top of the lake at spillway level. As a result, the lake was 188,000 acre-feet *below* its capacity. Flood gates were opened at noon on July 22, with the lake still 112,000 acre-feet below the "full" mark. But during a fourteen-hour period, 107,000 acre-feet of water was released while 163,000 acre-feet flowed into the reservoir.

The result was another Colorado rampage that produced severe damages all the way to the Gulf Coast. And the victims were enraged because they had thought that Buchanan Dam was supposed to give them permanent protection from floods. Rumors that the LCRA had failed to operate the dam properly spread even more quickly than the floodwaters had. Longtime opponents of the Authority seized upon the wrath of the river-area citizens and fanned the flames of protest, prompting Governor Allred to order an investigation of the matter. He called a mass meeting on July 30, 1938, at the State Capitol, and was greeted in the House of Representatives chamber by not only many elected local, state, and federal officials but also by about 200 farmers and businessmen from the flooded areas.

Fayette County Judge E. A. "Sam" Arnim was one of those who asked Allred to call the meeting. The original idea, he said, was to have three representatives from each of the five counties stretching from Bastrop to the Gulf Coast appear to discuss the situation with Allred and Senator Connally to "determine what could be done about the situation."

Arnim said they had asked the county judge of each county to see to it that three men would be at the meeting. "In doing that," he said, "we thought we were grabbing ahold of a yearling — and find it had grown into a full-sized bull overnight." He and his colleagues wanted to find out if the Buchanan reservoir was full prior to the flood. He asked:

> And if it was full, then why? We want to determine, through any means possible, whether or not that flood in the lower valley was due to negligence in the management or operation of the dam. Then, on a long-range matter, we want to know whether or not that dam is to be used exclusively for power purposes, or whether it is intended and will be used to give us at least a measure of flood control. I can assure you that the people living in the flooded area of the lower valley were under the impression that Buchanan would be used for a flood program. They are now convinced that it was not used for that purpose.

He and other speakers, most of them highly critical of the LCRA, demanded a thorough investigation and some even called for the "resignation or impeachment" of the Authority's board of directors.

Allred chaired the meeting and expressed confidence in his three appointees to the nine-member board. He also noted that the State Senate had a general investigating committee with powers of subpoena which could probe the matter.

Connally and Mansfield also spoke to the group before Johnson took his turn at bat. According to Roy J. McLean, an Austin court reporter who recorded the proceedings, Johnson said:

> Governor Allred, and distinguished guests: I am very happy to be here with Senator Connally and Congressman Mansfield, and those of you who have been affected by this destructive flood that has just visited us. I have heard a lot the last few days about the reasons for the destruction that has been visited upon us. If this is a man-made flood, if the project has been operated in such a manner as to contribute to this destruction, I want to know about it.
>
> While I realize the Colorado River Authority is a state

agency, created by you people to do certain work for you, those of us who have the pleasure of representing you in Congress have contributed materially to aiding the Colorado River Authority in the development that has taken place on the River.

When I went to Congress a little over a year ago, provision at that time had been made for three complete dams on the Colorado River. Buchanan Dam was just about completed; Inks Dam was underway; provision had been made for the Austin Dam; and five million dollars had been set aside for the Marshall Ford Dam. I had been there only a short time until the citizens of this area and of the lower valley impressed upon me the necessity of building additional structures, or raising those we had already begun, if we were to expect any flood control on this River.

With the aid of Congressman Mansfield, Senator Sheppard, Senator Connally and the entire Texas delegation, we secured a grant of five million dollars to add to the five million dollars we already had to erect Marshall Ford Dam to a height of one hundred ninety feet, in the hope that that would minimize the floods that frequently visit us. Engineers kept working on the flood control problems of the Colorado River, and only two or three months later came in and said, "It is going to take an additional two million dollars to build that dam to a height of one hundred ninety feet."

Because Judge Mansfield, in his wisdom and foresight, had passed through Congress legislation authorizing the Marshall Ford Dam project, Congress did not *lend* — but Congress outright appropriated two million and thirty thousand dollars to complete that dam to a height of one hundred ninety feet. Why? Because Congress realized that that was necessary in order to insure even an adequate amount of flood protection.

A few months passed and then the best engineers available to the federal government came in and said it was advisable, if we were to control such floods as we had in 1935 — and last week — to raise that Marshall Ford Dam from one hundred ninety feet to two hundred sixty-five feet. And just three weeks ago — to be exact, July ninth — with the aid of the people down the River, with the cooperation of all the people in this District, and with the help of those people you have selected to help you up there, President Roosevelt said: "I believe those engineers are right, and I don't believe we can control those floods with our present setup; and for that reason, I am taking one million two hundred and fifty thousand dollars from the appropriation set aside for federal projects, and we are going to extend the foundations of the present Marshall Ford already constructed — in order that, if experience dictates to us the necessity of doing it, we can go in there and by adding fifteen or sixteen million dollars more raise that dam to a height of two hundred sixty-five feet."

I feel that the flood that was with us, and is still with us, dictates to us the advisability of securing all of the control of that River that we can. And if the investigation that you provide for reveals that any engineer connected with your state Authority, or with the federal agency that has been financing this operation, has been negligent, I feel sure that you can feel confident that immediate action will be taken.

We mean to find out what mistakes were made, and to correct them. But we must not forget that we still have a big job to do on that River, and the little dams we have there are not ever going to provide for flood control entirely. But I feel like, with a group like this behind us, we can do almost anything; and I hope you people, when we are making our efforts to raise these dams to provide a maximum amount of flood control, will be enthusiastically behind us. I thank you.

Johnson's speech drew loud applause. Then, in response to questions, he went on to say that he understood that Buchanan Dam would provide some degree of flood protection but that the primary flood control dam would be the one at Marshall Ford. He noted that the Colorado River project was now complete, with the exception of Marshall Ford, and explained:

The Authority does not have the authority to acquire the land flooded by this large dam. The Legislature did not give it to them. No taxes have been remitted so the only recourse the Authority has is to say, if the federal government will put in twenty or thirty million dollars on this flood control at Marshall Ford, we, the State, will give you the land and you give us the dam. The federal government has yet to furnish about fifteen million dollars to finish Marshall Ford to two hundred and sixty-five feet; and whether the Authority is going to be able to secure enough funds, which will be about two million dollars of state money, to buy the lands that will be overflowed is dependent on their ability to sell enough water down the River and on power sales.

Wirtz, as general counsel for the LCRA, assured the group that the board wanted to construct and operate the project to produce the greatest possible benefits for the people.

"I think that any of you gentlemen who may know any of these Directors would know that the last thing any of them would do would be something that would bring destruction to somebody else's property," said Wirtz. He added that there would be no need to serve subpoenas on any of the LCRA's board or staff members because they would appear voluntarily for any investigation "and bring every scrap of evidence and records and subject themselves to cross-examination."

"We want to get the facts in just as bad as you do," he declared. "If anyone is to blame, they will not be shielded by the Board and, if any mistakes have been committed, we want to be sure they are corrected in the future."

Wirtz then presented a resolution which had been adopted the day before by the LCRA board. It said the board would welcome an investigation of all complaints to "determine whether blame should be placed on any agency or individual and, if so, to what extent."

That call was answered quickly. Moments before the meeting adjourned, State Senator L. J. Sulak of La Grange said he had just spoken by telephone to State Senator T. J. Holbrook, chairman of the Senate Investigating Committee, in Galveston. Holbrook, he announced, had called a meeting of his committee "and all interested parties" to meet in Austin on August 8 for a full-scale investigation. He added that the State Board of Water Engineers and the state reclamation engineer also were being asked to study the flood and present testimony to the committee.

While the meeting could not drown any sorrows inflicted by the most recent flood, it gave a new sense of urgency to the long-sought goal of taming the Colorado.

12

High
Time

Ironically, the investigation of the flood controversy came just as the LCRA was launching an intensive campaign to find markets for its electricity. The timing compounded both problems.

The Colorado River Improvement Association, headed by Dr. Goodall Wooten of Austin, went to bat once again for the Authority. The association, which had promoted river development for nearly half a century, organized for what its directors hoped would be their last battle for flood control.

Mass meetings were held in Wharton, Burnet, Lockhart, Georgetown, and Austin. Several newspapers in the Colorado Valley published editorials defending the LCRA. The *Austin American*, for example, asked how municipal and irrigation needs could be met in times of drouth if the reservoirs were kept empty to catch floods; it also reminded people that hydroelectric power sales were necessary in order to pay off the indebtedness on the dams.

The *Dallas Morning News*, however, in an editorial on August 1, 1938, used the flood operations of the LCRA to attack the Authority's hydroelectric sales program. But, strangely enough, on the same day, the *El Paso Herald-Post* — even though far removed geo-

119

graphically from the Colorado and its controversy — came to the defense of the LCRA. That editorial said:

> Since its birth four years ago, the Colorado River Authority has been accused of every "crime" a public agency could commit but it will never be convicted of more than one — that of offending the Texas power trust by proposing to sell cheap electricity to municipalities.
>
> The latest "crime" laid at the door of the LCRA is causing a flood.
>
> Since the CRA in no sense can be charged with making an unprecedented cloudburst fall around San Saba, the charge is made that the Authority could have stopped the flood if it had not been contemplating making electricity.
>
> The flood control system of the CRA is only half complete, and nobody ever said that Buchanan Dam, the only flood control dam now finished, could control floods all alone. If it could, there would be no need for Marshall Ford Dam.
>
> Buchanan, far from contributing to the flood, held back enough flood water to cover 300,000 acres one foot deep. It justified its existence, and the fact that it did not stop the flood only proves something that engineers already knew, that Marshall Ford is vital to flood control.
>
> Then why all the hullabloo against the CRA?
>
> It is just another attempt to wreck the Authority, and to prevent it from demonstrating, through a power yard-stick, that rates charged for electricity in Texas are too high . . .
>
> We are glad there is going to be an investigation. We hope it is thorough. That it will result in clearing away all the clouds that have been raised because the Colorado Valley was unfortunate in having its greatest flood before the flood control system was completed.
>
> It also will clear away the clouds of propaganda thrown on the project by selfish business interests and their apologists.

Holbrook's Senate Investigating Committee began its probe on August 8, 1938, with only three of its five members present: Holbrook and Senators Wilbourne B. Collie of Eastland and Albert Stone of Brenham. Senators R. A. Weinert of Seguin and Joe Hill of Henderson missed the spirited debate, which eventually stretched into 1939.

Among the first witnesses was T. U. Taylor, the retired dean of engineering at The University of Texas. He expressed the opinion that the flood could have been averted if the Buchanan Dam gates had been opened forty-eight hours earlier and insisted that unless Lake Buchanan was kept empty, the flood would be repeated. Said Taylor,

> To have flood protection, a dam must have its gates open and its lake full of air. As a reservoir for power production it must have its reservoir full of water. The two functions — power production and flood control — are absolute antagonists. The choice must be made between the farmer in the Valley or the money dividend in the hills. This disaster indicates that someone in the River Authority should be elected to a life membership in the "Bonehead Club."

But an entirely different view came from Harry W. Bashore, a U.S. Bureau of Reclamation engineer who had been sent to Austin by Secretary Ickes to study the causes of the flood. Bashore described the huge quantities of water that roared down the Colorado during the flood and estimated that Buchanan Dam had managed to reduce the level of flooding at Austin by one foot. He concluded that the flood proved primarily the need to complete a higher dam than had been planned at Marshall Ford.

After the Senate committee recessed until August 17, Austin's civic leaders decided to stage a rally in defense of the LCRA on August 16 at the Travis County Courthouse. According to the *Austin Statesman*, Mayor Tom Miller told an overflow crowd of 400 at the rally that much of the criticism aimed at the agency had been prompted by the private power companies who were "jealous of the CRA's plans for production of electrical power from some of the dams." The *Statesman* article, by Morris Midkiff, said:

> Mayor Miller charged there is a movement on foot to cripple the Colorado River Authority by sponsoring legislation to prevent it from making and selling power; and he stressed the importance of the completion of the dams program to Austin.
>
> He pointed out that a federal engineer "in the civil service since 1906, and who has served under several Republican administrations" had made a survey for Secy. Ickes and made "an impartial report refuting charges" that the Colorado's July flood was "man made."
>
> "For the greatest good of the greatest number the dams must be completed," Mayor Miller declared. "And the dams must be paid for in some manner. The state cannot pay for them. It will be too busy trying to find revenue — some $40,000,000, to pay

O'Daniel's pensions to everyone over 65 — to find any revenue for paying for the dams. So the only way the loans for dam construction can be paid back is by manufacturing and selling power . . ."

W. Lee (Pappy) O'Daniel, a radio announcer who had gained enormous popularity while selling flour with the aid of a western band, had won the Democratic nomination for governor on a "platform" of the Ten Commandments plus a pension for everyone over sixty-five. His election, in those days of one-party Texas government, was assured. He defeated Republican nominee Alexander Boynton in November with 473,526 votes to 10,940.

Mayor Miller's speech was mild compared with the vitriolic attack by Congressman Johnson on the LCRA's foes. Johnson told the rally:

> Yes, we are going to have four dams. They are going to hold back flood water and they are going to pay for themselves with some electric power which doesn't have to run through the cash register of a New York power and light company before it gets to our lamps . . .
>
> The time has come for all citizens of Austin and for every citizen of this part of Texas to get a hold of his representative in the Legislature and to say to that representative:
>
> "We are going to have dams on the Colorado River to control floods. We are going to complete the four dams we are building now, and we are going to build some more dams if we find they are needed to do the job.
>
> "We are going to keep building these dams in a business way. When we store up flood waters we are going to release them through hydro-electric turbines and we are going to sell the electricity those turbines make to the people. It will be the people's electricity and the people are going to get it at cost — for a small fraction of what they have been paying the power monopoly for twenty years . . ."
>
> As far as I am concerned, we are going to say to *The Dallas News* and the TP and L, "We are going to build our dams and we are going to keep our men at work," and that is what I want you to join me in saying tonight . . .

The Texas Board of Water Engineers also investigated the flood and its causes, then reported to the Senate committee, on September 19, that it felt the criticism of the LCRA was unwarranted. The engineers noted that the flood was the largest of record which had ever occurred above Buchanan Dam, that the Buchanan was not large enough to control such floods, and that the flood on the Colorado River below Austin was *not* "man-made." In order to prevent a "man-made" flood, said the board, adequate information should be made available and studied carefully before the release of water from the reservoir in advance of a flood.

State Reclamation Engineer R. J. McMahon also had made a study, at the Senate committee's request, and reported similar findings the same day. He said Buchanan Dam had actually retarded about 200,000 acre-feet of floodwaters.

"Of course, had the Colorado River Authority had storm data in advance, the gates of Buchanan Dam might have been better-regulated at the beginning of the flood," said McMahon. His theory was that the peak flood of the Llano River at the time it came into the Colorado River, in addition to a large inflow into the lake, caused Austin's flood.

He estimated annual losses from floods on the lower Colorado at about $5 million and said records dating back to 1891 showed that twelve storms out of twenty-three had occurred along the Balcones Fault line below the Marshall Ford Dam location.

"Therefore," said McMahon's report, "it is recommended that the high Marshall Ford Dam be completed with over 3,000,000 acre-feet of total storage. The parallel levee system from Columbus to Matagorda should be built. A report on the design of this system has just been released by the Bureau of Reclamation showing an estimated cost of $3,000,000."

McMahon further recommended that a complete system of rainfall and river gauges be installed in order to give advance warning of floods.

The LCRA already had taken action on the latter recommendation. On September 6, 1938, the board approved an agreement with the U.S. Weather Bureau under which the LCRA would install fifty rain gauges in the upper watershed of the Colorado and its principal tributaries to supplement the approximately forty Weather Bureau gauges then being used in that area. Rain-gauge observers were employed to make daily readings and report directly to the LCRA's Austin office so the staff could be quickly and fully informed about any storm or rainfall capable of producing major flood conditions.

This was the beginning of the first comprehensive watershed

reporting system in Texas, and some of the rain gauges were installed outside the LCRA's district. The full system, with gauges to measure the volume of water in major streams feeding the Colorado as well as in the river itself, was not completed until March of 1939. At that time, the LCRA contracted with the Board of Water Engineers and the U.S. Geological Survey to install twelve stream-flow gauges, equipped with shortwave radio transmitters, to broadcast information on stream stages automatically.

The LCRA's own investigation of the flood and its causes had shown clearly that operating personnel had been handicapped by the lack of accurate information as to the volume and duration of rainfall on the San Saba, Llano, and Pedernales river watersheds.

Early in 1939, the Senate Investigating Committee issued its report and said it found that the dams were being operated primarily to produce revenue from the sale of power in order to amortize the bonded indebtedness over a period of thirty years. Flood protection, said the report, was to be merely incidental to power production. And it added: "The question as to the primary purpose of such projects in Texas is one on which this State has not yet declared a definite public policy."

Senator Sulak then introduced a bill that would require such public authorities as the LCRA to retain in normal times a maximum of only fifty percent of the water which could be stored in their reservoirs, reserving the other fifty percent of capacity for flood protection. He withdrew the measure after the LCRA directors announced that they would reserve 800,000 acre-feet of storage capacity at Marshall Ford exclusively for flood control purposes.

The board had before it at that time a report from U.S. Bureau of Reclamation Engineers E. B. Debler and John R. Ritter recommending that plans for the Marshall Ford Dam be changed so that it could be built higher. The 270-foot height they proposed for the dam would provide approximately 800,000 acre-feet for flood control storage.

The board instructed McDonough and Wirtz to work with Congressman Johnson and the Bureau of Reclamation on the project. They were urged to push for the high dam.

Wirtz later wrote a memo to the LCRA board noting that "parties interested adversely to LCRA had employed private counsel headed by former Governor Dan Moody" to lead the Senate committee investigation. He elaborated:

While the ostensible purpose of this investigation was to determine whether or not the Authority was negligent in connection with the flood, interests adverse to the Authority were quick to take advantage of the situation, and the investigation developed into an inquiry into all phases of the Authority's program with the idea of determining whether there should be legislative control of the operation of the Authority's facilities, with the idea of curtailing the production of electrical energy and diminution of its revenues to such an extent that it would be impossible to meet its obligations.

If Wirtz's theory was correct, the conspiracy against the LCRA apparently backfired. The probe convinced a lot of people that a "lake full of air" to prevent floods could be produced only by building a higher dam at Marshall Ford.

13

Help
Wanted

The Senate committee's investigation did little to dilute the LCRA board's headlong plunge — under almost constant prodding from Congressman Johnson — to develop markets for electricity and establish a system for delivering it. After LCRA officials conferred with them, the governing bodies of many small towns and cities in the area decided they wanted to buy their power from the LCRA; they began exploring ways to either purchase the privately owned electrical systems which served them or to build their own systems to compete with those.

Meanwhile, the LCRA also was negotiating with two of the privately owned power companies — Texas Power and Light Company and Central Power and Light Company — on the possibility of buying the transmission systems which they owned within the LCRA trade area. And, at the same time, the relentless efforts to obtain more federal aid continued in Washington.

The summer of 1938 turned out to be long and hot, but it opened the floodgates of opportunity for the LCRA. Efforts had been initiated during the spring to take advantage of the possibilities opened up by the creation of the federal Rural Electrification

Administration in 1935 and the Texas legislature's subsequent passage, in 1936, of the Texas Electric Cooperatives Act.

Most of the small towns in the LCRA's market area had at least some form of electricity when the agency was born. A few were fortunate enough to have Texas Power and Light service, but in many towns the only power supply came from nothing more than a thirty-horsepower diesel engine which provided energy for ten-watt light bulbs and some small motors. In the larger towns, more appliances and motors generally could be used; however, in most of those the electric plants usually shut down at midnight.

Even the few conveniences made possible by such limited power services in the towns were virtually unknown on the farms and ranches of Central Texas. As late as 1935, according to the U.S. Department of Agriculture, no more than ten percent of the farms and ranches in the United States had access to central station electric service. Texas ranked first in the number of farms but forty-fifth in the number of farms receiving central station electricity. Only 2.3 percent of the farms in Texas were electrified, and practically none of those was in the Hill Country. Throughout the state, the electric lines extended only short distances beyond town limits.

Thus, while urban dwellers began getting helping hands from electricity, rural folks still performed their chores in primitive fashion. Farmers and ranchers who lived only short distances from electric power lines were told by the utility companies that it would cost too much money to hook their houses up to the lines. For example, the Texas Power and Light Company's policy permitted only farms within fifty yards of a line to be served by it. Hundreds of rural homes were near such power sources and many, just barely outside the fifty-yard limit, were refused service. Even people who offered to pay the entire cost of extending the lines into their homes were refused because TP&L said it did "not wish to set a precedent."

Thus, farm wives who knew that their city sisters were using those wonderful electrical appliances had to continue drawing water by hand from wells, cooking on woodstoves, depending on kerosene lamps for light, and using outhouses instead of indoor plumbing.

For almost thirty years, the privately owned electric power monopolies throughout the nation had fended off efforts to have the federal government force them into supplying electricity to farms and ranches. Despite arguments that much of their electricity was generated by water power on rivers belonging to all the people, and not just to the power companies, they were able to keep hydroelectric power sources under their control until Franklin D. Roosevelt became president. Roosevelt insisted that river-generated electricity should be made available in rural as well as urban areas — making him an instant hero in such places as the Texas Hill Country.

First came the Tennessee Valley Authority, with its massive dams and hydroelectric plants. Next, in the Pacific Northwest, the Bonneville Power Administration on the Columbia River began providing electricity to rural areas.

On May 11, 1935, Roosevelt signed an executive order creating the Rural Electrification Administration (REA) as a part of his federal relief program. The new agency's lack of progress prompted his administration to introduce a bill in January 1936 granting the REA a certain amount of independence but making it a part of the U.S. Department of Agriculture. The measure would authorize REA loans to farmer-owned cooperatives and also to individual farmers in order to make electricity available in rural areas. The bill won quick passage in the U.S. Senate but encountered strong opposition in the House of Representatives, where the utility lobby exerted a great deal more influence. In the end, it was Sam Rayburn who rescued it from what appeared to be a terminal illness.

Working behind the scenes, "Mister Sam" helped engineer a compromise which made privately owned electric utilities also eligible for the federal loans to finance rural line construction. But preference was to be given to "states, territories and sub-divisions thereof, peoples' utility districts and cooperatives, and non-profit or limited dividend associations."

On the floor of the House, Rayburn made a dramatic plea on behalf of the farmers and their wives who needed electricity but were being denied it by the private utilities which claimed it would not be profitable. Rayburn cited Texas as an example and declared: "When free enterprise had the opportunity to electrify farm homes — after fifty years, they had electrified three percent!"

Roosevelt signed the bill into law on May 11, 1936 — exactly one year after issuing the executive order which had established the REA. That, coupled with the Texas Electric Cooperatives Act, provided the legal basis for the organization of rural electrification cooperatives, which could obtain loans from the REA. But residents of the Hill Country still faced an uphill battle because the REA had two major requirements for making loans: (1) the coop receiving a loan had to offer evidence that it could repay the principal, plus three percent interest, within twenty-five years; and (2) the lines to be built had to serve an average of at least three farms per mile.

That "three farms per mile" requirement seemed an insurmountable obstacle in the wide open spaces of the Hill Country. But Congressman Johnson did not seem to believe in such things as "insurmountable obstacles," even after REA Administrator Morris Cook visited Austin and threw cold water on the hopes of rural leaders from Burnet, Blanco, Hays, and Llano counties. Cook told them they had little chance for success because they simply had too much land and too few people. The cost of extending the lines would be so great, he said, that they would be unable to develop enough revenue to repay the loans. And he said the REA would *not* waive the "three customers to a mile" criteria for a loan.

One of those attending that meeting was E. Babe Smith, a rancher who had graduated from Southwestern University in Georgetown and lived in northeastern Burnet County, an area known as "The Dark Corner" of Burnet County. In 1936 he had heard of efforts to organize an REA coop in Bell County and extend its lines to Bartlett, only forty miles from his ranch. When he checked into the possibility of having those lines stretched to his ranch, he was told there was no hope because of the sparse population between it and Bartlett.

Thus, when he was asked to attend a meeting in Johnson City concerning the organization of a Pedernales Electric Co-operative (PEC) Smith said there was no use in his going; his ranch was just too far away to get electricity. But at the insistence of Roy Fry, he finally agreed to meet with Fry and Congressman Johnson in Fry's drug store at Burnet.

Smith appreciated more than most of the others in that area the value of electricity, since he had lived in Lampasas and had seen the benefits of service provided there by the Texas Power and Light Company — usually only from sunset until 10:00 P.M. Even that limited power supply showed him how much electrical appliances could reduce the housework burden for his mother.

As much as he wanted electricity, Smith thought it was a hopeless cause so far as his ranch was concerned. But when he told Congressman Johnson he felt it would be impossible to get an REA loan for line construction in that area, Johnson told him: "I'll get the money! I'll go to the President if I have to, but we'll get that REA money!"

During a 1983 interview, Smith recalled that drug store conference and said Johnson had not only inspired him but also "made me feel there was a chance."

As a result, Smith became one of the most ardent recruiters of Pedernales Electric Co-operative members. The PEC was organized on May 8, 1938, in Johnson City, and its next meeting was held there on June 20. About sixty people attended that one, Smith recalled, and many of them either were not in favor of trying to get an REA loan or felt it would be impossible because of the "three to a mile" requirement.

Congressman Johnson explained that each person who wanted electricity would have to sign up as a member and pay a $5 deposit, which would also serve as a membership fee. He told them that, collectively, they would be responsible for retiring any loans made by the federal government — and that they would be required to grant easements across their land not only for their own lines but also for those needed to serve their neighbors.

Henry Grote, the Gillespie county agent, was among those most antagonistic to the idea, according to Smith. Grote told Johnson that he was trying to talk the farmers into something that might cause them to lose their farms if they were unable to pay their electric bills. Grote and others felt that the potential benefits of electricity were not worth the risk involved.

"To them, it was only a light bulb," said Smith. "And in those days, there were a lot of people who just didn't have the five dollars."

In addition, he noted, most of the people knew so little about electricity that they had a deathly fear of it. They had heard that birds sitting on the electric lines would be killed; some said that the lines would attract lightning bolts, which would kill the cattle grazing under them or burn down houses and barns. Many felt they did not need electric lights, since kerosene lanterns had served them well for years.

Johnson tried to explain all the things that electricity could do, including its ability to run farm equipment. He inspired people with the belief that if they could get most of the people signed up, three to a mile or not, he could get the REA loan. As a result, a lot of people went away from that meeting determined to sign up their neighbors. But their task was not easy.

In the final analysis, said Smith, it was the housewives of the area who brought about success. They wanted relief from the drudgery of pumping and carrying water for cooking, cleaning and bathing, from carrying wood for cooking and heating, and from having little or no hot water. And most of all, he said, they wanted inside bathrooms.

Smith recalled one meeting attended by about 100 farmers and their families in the Kempner schoolhouse. As Caro described it in *The Path to Power:*

> At the beginning of the meeting, Smith passed out PEC applications; only two of the farmers signed them. So Johnson tried *his* hand at persuasion. "He played on the emotions of the women," Smith says. He talked about his mother, and how he had watched her hauling buckets of water from the river, and rubbing her knuckles off on the scrub board. Electricity could help them pump their water and wash their clothes, he said. When they got refrigerators, they would no longer have to "start fresh every morning" with the cooking. "You'll look younger at forty than your mother," he told them. Because he was the congressman — and, to rural people, therefore an object almost of awe — and because he was "very persuasive," they listened to him attentively. But they didn't sign up. On the Fourth of July, 1938, he drove with Smith from picnic to picnic — Smith had brought along a Sears, Roebuck catalogue to show them the washing machines and refrigerators they could use if they had electricity — but collected only a few handfuls of signed applications. "In the car going to the next town, he [Johnson] would say: 'What's the matter with these damned people? You offer them something . . .' He'd get discouraged." Despite months of effort, the county agents had collected nowhere near the number required — not three per mile, but fewer than two. The Pedernales Electric Co-op appeared stillborn. Johnson warned the farmers that if they didn't establish a cooperative to purchase the electricity created by the dams, TP&L might buy it — and offer it only at rates they could never afford. In a speech entitled "Our South," he said, "I believe we should use that power. I believe that river is yours, and the power it can generate belongs to you."

The "salesmen" for the LCRA and the PEC, including Smith and Johnson, fanned out across the area trying to sign up customers for electricity. Eventually, they were able to sign up just barely more than two connections per mile. It was then that Johnson went to President Roosevelt and persuaded him to have the REA make an exception to the "three customers per mile" rule.

Johnson used his friendship with White House aide Thomas G. (Tommy the Cork) Corcoran to gain an appointment with the president. And while he was in the inner sanctum, Johnson said later, President Roosevelt picked up the phone and called REA Administrator John Carmody. The president told Carmody, according to Johnson, to make the exception and said, "Those folks will catch up to that density problem because they breed pretty fast."

As a result, the Pedernales Electric Co-op received a telegram on September 27, 1938, declaring that the REA had made it a loan of $1,322,000 (the amount later was raised to $1.8 million) for the construction of 1,830 miles of electric lines which would bring electricity to 2,892 families in the Hill Country.

Judge Ferguson recalls meeting a man several years later who had been a lawyer for the REA at that time but had moved on to a trade association job.

> He told me about Lyndon and Pedernales Electric. He said they got the application for the Pedernales Electric Co-op and they had two stacks of applications. One was a stack of those the White House had not indicated any interest in and it usually took about six months to work down that stack. They had another stack that the White House *had* indicated some interest in and it took about three months to work down that stack.
>
> He said that one morning the White House called and said Congressman Johnson was interested in the Pedernales Electric Coop's application and they just wanted to let them know the White House also was interested in it. So, he said, they took the Pedernales application out of the six-months stack at that point and put it in the three-months stack.
>
> And he said the phone rang directly and it was Johnson. He asked when the papers would be ready on the Pedernales Coop. This lawyer told him it would be "about three months or something like that." Johnson said, "Three months, hell! I'm going to take them down there *tonight!*" They tried to assure him they'd get right on it but it just couldn't be done in a day's time. But Johnson said, "I want to take those papers to Johnson City tonight; I'm going to fly down there." And when he hung up, it wasn't but a few minutes until the phone rang again. This time, it

was the White House and they said, "Congressman Johnson wants to take those papers to Johnson City tonight — and the White House would be very grateful if you'd get them ready."

And this lawyer told me that, "believe it or not, we got 'em ready!"

Meanwhile, the LCRA board had taken a giant leap in its campaign to sell electricity — and, once again, it did so at the urging of Lyndon Johnson.

McDonough was highly respected as a construction expert but had made it clear to the board, when he was hired in 1935, that he wanted to stay only until the initial dams, power plants, and other facilities had been built. As completion of the Buchanan and Inks power plants neared, he was among those who recommended the hiring of an operations manager to take over the direction of all LCRA facilities and also the program to sell power.

But it was the not-so-gentle persuasion of Congressman Johnson which prompted the board to take what turned out to be one of the most important steps in its history. In a memorandum dated July 19, 1938, Johnson told the board:

> I. It is not only my sincere conviction that the Power Committee of the Lower Colorado River Authority should operate as a thoroughly cohesive, militantly efficient unit, but that it should, from this point forward, have a place for the direct delegation of its powers and authority; a representative upon whom it likewise may affix direct responsibility.
>
> II. This Power Representative of the LCRA should be selected on the basis of his ability to solidify unorganized forces, plan judicious forms of attack, disseminate understanding of the whole complex program, and promote and sell the one commodity the Authority has to dispense.
>
> III. He should be able, as the prime requisite, to find and expand centers of co-operative feeling. He should be thoroughly aware of the fact that he is a pioneer and that he is leading a crusade toward a revolutionary step in Texas social and economic life.
>
> IV. He should plan his program on the following general lines:
>
>   a.   By direct personal work in the field.
>
>   b.   By contact with every city official in the area, and with every actual and prospective leader in rural areas.

c.   By giving assistance to cities in election fights for public distribution of power.

d.   By acting as chief source of inspiration and information in every campaign for the financing and establishment of public distribution systems for power.

e.   By knowing the answer to every question which can be brought up in regard to setting elections, signing of contracts, and reaching the proper Federal agencies for loans and grants.

V. He should be the counterpart of the General Counsel of the CRA, its Engineering Authorities, its Construction experts, and its Representatives in Congress, in his own field.

VI. He should know the entire history of the growth and development of Private Utility Management and should be thoroughly versed in the growth and progress of the movement for effective public ownership and distribution.

VII. By newspaper and magazine articles, by radio and by speeches to every sort of organization he should be able to present the story of the present Federal power program in such a manner that there can be no possible mistake about it.

VIII. He should have at his fingertips *the whole picture* of the entire CRA potential field. There should not be a square mile he does not know — with all the ramifications of its problems.

IX. He should be a man who will not only exhaust every diplomatic means toward the pacific solution of problems, but one who will never swerve from his purpose in so doing and who, when forced to the extremity of duplicating lines and entering direct competition, will prove entirely relentless.

X. He should be given a free rein, solid backing and assistance, and from him the committee should be able to expect immediate and direct results and progressive accomplishment until the last available supply of public power is disposed of.

XI. He should be freed of all entangling alliances with present departments but, through the offices of the Power Committee, should have the benefit of the co-operation of every department of the CRA. He should likewise, through the committee to which he is answerable, co-operate fully with every other department.

Lyndon B. Johnson
Member of Congress
Tenth District of Texas

Austin, Texas
July the nineteenth
1938

All that was needed, Johnson seemed to be saying, was a Superman. It might have taken the directors a long time to find the right man for the job, had the idea come to them as a complete surprise. But Wirtz had the ideal candidate standing in the wings and already had persuaded him to take the post when the board met on August 2, 1938.

At that meeting, the directors voted to create the position of operations manager. They also approved unanimously the recommendation of their Power Committee to hire Max Starcke, then mayor of Seguin, for the job. His salary was set at $10,000 per year and his term would run from the next day, August 3, 1938, until January 1, 1942.

There was nothing in the chubby, forty-nine-year-old entrepreneur's appearance that came even close to resembling Superman. But Starcke turned out to be a super salesman and an outstanding organizer. He was destined to become the most beloved general manager of the LCRA's first half century.

14

The Seguin Song

In Max Hugo Starcke, the LCRA got a livewire salesman-pro-moter-diplomat-administrator who eventually played an active role in fifty-six different civic, social, and professional organizations. He may be the only man in history who organized a Lions Club and later founded a Rotary Club — without losing any friends among the Lions.

Starcke was born November 11, 1884, on a farm near Seguin, in the community of York's Creek. When asked where York's Creek was located, he used to explain with a grin that it was "near Zorn and Geronimo." His family moved to Seguin when he was about four years old, and he graduated from high school there.

He attended Texas A&M for two years and then transferred to a business school in San Antonio, paying his way by working as a salesman for some coal mines near Uvalde. In 1906, at the age of twenty-one, he became a law clerk and secretary to Texas Supreme Court Justice J. B. Dibrell. When Judge Dibrell bought some citrus land in Jim Wells County that included the tiny village of Sandia, he hired Starcke to lay out the streets there, subdivide some of the land, and develop the town.

While Starcke was handling that chore, he was approached by

a couple of neighboring ranchers who told him they would like to sell off part of their land. Starcke contacted Dibrell and, as a result, the two of them made a deal which resulted in their establishing the town of Orange Grove.

According to *Fellow Texans in Profile,* a book published in 1948 by the editorial staff of the *Austin American-Statesman,* Starcke could not wait to get back to Seguin, despite his success as a developer. He moved back home and opened a livery stable, which became quite successful. He then acquired a hearse and established a funeral home, building up assets that would later enable him to start a bank, and when he was only twenty-six years old he was elected an alderman. Some of his friends insisted later that he run for mayor of Seguin, and his first campaign for that office was the only hard one he ever had.

Sim Gideon said he heard later that when Starcke first ran for mayor, the City of Seguin was getting its water from the Guadalupe River "and they said it was so muddy you couldn't see your toes when you were taking a bath." Starcke made a unique campaign promise: that if he were elected, within a year the people would be able to see their toes in their bath tubs.

He not only cleaned up the water; he also made arrangements for a bountiful supply of it. And he was elected mayor for six consecutive two-year terms.

Starcke built Seguin's first water filtration plant and helped acquire land for the six dams on the Guadalupe River in Guadalupe and Gonzales counties. He built a hydroelectric power plant on the Guadalupe River and brought its benefits home to the people by enabling them to use all the water they wanted for a *maximum* charge of only $1.50 per month. Seguin quickly became a city of lush lawns and beautiful flowers.

Starcke paved the city's streets, cleaned up many eye-sore creeks throughout Seguin, and built parks and playgrounds for all segments of the population. The City Council was so grateful for his work in establishing a show-place park — which included a golf course, swimming pool, picnic facilities, and other amenities — that they named it Max Starcke Park. He gained widespread attention for building, without raising taxes or borrowing any money, the first air-conditioned city hall in Texas *(Fellow Texans in Profile).*

Meanwhile, he organized the Farmers State Bank of Seguin and directed its operations for fifteen years. In his spare time, he also organized a city baseball team and served as its manager.

When he became mayor and learned that little information on municipal matters was available to city officials, he helped organize the Texas League of Municipalities (later renamed the Texas Municipal League). Starcke served as its president for several terms and as president of the South Texas Chamber of Commerce for three years.

His remarkable record plus his popularity with other city officials, especially those in the LCRA area, made him an ideal choice for the operations manager's position.

With two dams, Buchanan and Inks, then complete and two more, Marshall Ford (later renamed Mansfield) and Tom Miller, under construction, he went to work immediately developing a market for electricity.

Proceeds from the city-owned utilities had made Seguin practically a tax-free community, an example he used to good advantage in promoting municipal ownership of utility systems throughout the LCRA area. "Look what it did for Seguin" became his theme song.

Two weeks after taking office, Starcke sent a questionnaire to thirty-two Texas cities that owned and operated their own electric power systems. He wanted information on their successes and failures. Of the twenty-six that replied, not even one reported a loss on operations and most had shown large profits. Georgetown, for example, had a net profit of $24,624 during the 1937 fiscal year, even after furnishing free electricity worth $6,000 to city departments.

The survey results were compiled and published, along with information on the availability of federal loans and grants for building publicly owned electrical distribution systems, and sent to city officials in the LCRA area. The same packet, which went out in early September of 1938, gave details of the Seguin success story as an example of what small cities could expect if they established their own systems and bought electricity at wholesale rates from the LCRA.

Almost immediately, city councils in the area began adopting resolutions favoring municipally owned power systems and, in most communities, they drew support from the local newspapers. During September, October, and November, ten cities held bond elections to raise money for building or purchasing their own power systems.

Texas Power and Light, along with other private utilities operating in the area, fought fiercely against the proposals. Just before each election, TP&L paid its employees in silver dollars to impress local merchants with the size of its payroll. Starcke countered those efforts by giving his own testimony on how Seguin had made so many municipal improvements without raising taxes.

The bond issues were approved in all ten cities: Blanco, Burnet, Elgin, Fredericksburg, Gonzales, Kyle, Lampasas, Llano, Moulton, and Schulenberg. Only in Gonzales, which was served by Central Power and Light, was the vote even close (302 for, 274 against). In most cases, the people voted overwhelmingly in favor of municipal ownership — assuring the LCRA of a market for its electricity and enhancing its bargaining position in trying to buy existing distribution facilities from private firms.

Four private power company executives had met with the LCRA board in early August to discuss the inevitable conflict. They were John W. Carpenter, president; William Lynch, vice-president; and Joe Worsham, attorney (all of the Texas Power and Light Company); and E. B. Nieswanger, president of the Central Power and Light Company.

They tried to persuade the board to abandon the idea of selling directly to municipalities and rural coops and, instead, to sell them all the power which would be generated at the dams. The board steadfastly refused, however, to change its policy and made it clear that it intended to make electricity available to the people of Central Texas as cheaply as possible. At a second conference later in the month, the board offered to purchase the transmission lines and distribution facilities owned by the two companies in the LCRA market area.

On September 6, 1938, Carpenter presented a letter to the board offering to recommend that his stockholders sell all their facilities in the LCRA area — in sixteen counties — for $7.3 million *if* the LCRA would agree not to sell electricity in any of the other counties served by the firm. But Carpenter committed a major error in protocol by announcing the proposed terms to the press before sending them to the LCRA board. The LCRA directors were as infuriated as Congressman Johnson, who took such offense at the tone of Carpenter's letter that he immediately told him to "go to hell."

Later, the cool-headed Wirtz called Johnson aside and assured him: "Just because you tell a man to go to hell doesn't mean he has to go."

Despite the negative reaction of the board, the LCRA staff made a thorough survey of the TP&L properties in the sixteen counties involved. After looking over that report on November 21, the board approved a letter from Starcke to Carpenter formally rejecting the company's offer and entered it, in full, in the minutes. Starcke's letter pointed out that

> we have had executives and engineers of the Authority make every effort to determine the fair value of your properties. We were informed by you that your cash investment in electric properties amounts to $6,584,577 and in ice and water properties, $813,439, making a total of $7,395,016. We learned later that your books did not accurately show the investment in these particular properties inasmuch as they had not been segregated on your books, and the sum you named was an estimate.

Starcke declared that Carpenter knew, when he made the offer, that LCRA had no legal right to acquire and operate TP&L's ice-making properties and that the matter of supplying *treated* water to the cities was not connected with its business. But that was a relatively minor sticking point in his letter.

> It appears that the properties which you priced to us at over $7,000,000 are assessed in the counties for only about $1,700,000. On the face of these figures, it would appear that your cost price was set up for the purpose of making high rates and that it does not have the proper relation to the cost you set up for the payment of low taxes.

He said the LCRA's engineers had advised him that the TP&L facilities which Carpenter offered could be duplicated for approximately $4.6 million and that a twenty percent depreciation charge also should be applied. He continued:

> Therefore, because of the depreciation and obsolescence of various items of equipment, the property now is worth only 80 percent of what it would be worth new, making its value at the present time approximately $3,675,000. We are further advised by our engineers that, based on the Authority's present construction costs, it would be able to duplicate the electric service the Company now is rendering at an even less cost.

Starcke added that, in an effort to avoid competition, the LCRA was prepared to pay $4 million for TP&L's electric properties, even though that amount was approximately three times the assessed valuation and was more than it would cost to duplicate the facilities. He also took exception to a provision in Carpenter's proposal which would have required the LCRA to confine its activities to the sixteen counties involved: San Saba, Lampasas, Burnet,

Llano, Kerr, Blanco, Hays, Guadalupe, Caldwell, Travis, Bastrop, Lee, Washington, Austin, Fayette, and Colorado.

The LCRA, Starcke declared, would not take any action

> which would in any wise hinder this Authority in its purpose to cooperate with other authorities in the development of other rivers and in making available to cities cheap power without the intervention of private profit.
>
> The fact that you have been earning and expect to continue to earn a high return on the values which you claim to have been invested in these properties is sufficient proof of statements made heretofore by the Authority that the consumers' rates can be reduced from thirty to fifty percent below the rates now being charged by your company. Because we are handling public funds and are charged with a public trust, we are unable and unwilling to pay the exhorbitant price which you have asked for your properties, and inasmuch as you have indicated that you are unwilling to sell at the fair price we offered you, this will serve as a formal rejection of your offer.

It did not take Carpenter long to respond. He telegraphed a counter-offer to sell the TP&L transmission and distribution facilities within the sixteen counties — plus six hydroelectric generating stations and small dams in Marble Falls, Llano, Kerr County, San Marcos, Alvord, and Martindale — for $5 million.

The LCRA board accepted that offer on January 2, 1939, but the sale was not closed until September 1.

Meanwhile, Congressman Johnson was meeting with many farm and ranch leaders throughout his Tenth Congressional District, urging them to join hands and form a single rural electrification coop to work with the LCRA in using TP&L service lines for widespread distribution of electricity.

But he encountered opposition from the outset, not only in his own Hill Country area but also in the southern part of the district. There, rural leaders from Lee, Washington, Bastrop, Burleson, and Caldwell counties had been working for some time to organize a coop for that area. And even as the Pedernales Co-op was being formed, many people in Gillespie, Kerr, Llano, and other counties expressed more interest in forming their own coop than in joining with Blanco, Burnet, Hays, Travis, and other counties. These people also were unhappy with Johnson's plan to locate the headquarters in Johnson City; they preferred Fredericksburg.

This friction planted the seeds for what would eventually blossom into the "Texland" controversy, which was to prove a thorn in the LCRA's side for many years. The Pedernales Electric Co-op finally split into two groups, with the main offices of the PEC located in Johnson City while those of the newly formed Central Texas Electric Cooperative went to Fredericksburg.

The same day the LCRA board agreed to buy the TP&L properties, it applied to the federal Reconstruction Finance Corporation for a loan to pay for them. And it also began to press its negotiations with the Central Power and Light Company for similar arrangements involving nine cities it served in the LCRA market area. The directors even suggested that the LCRA could rent CP&L transmission lines as an alternative to outright purchase.

The LCRA was dealing from a position of some strength because the nine municipalities (El Campo, Cuero, Gonzales, Luling, Waelder, Moulton, Schulenberg, Fredericksburg, and Edna) all had voted to operate their own electric distribution systems and buy their power from the LCRA. But CP&L continued to oppose the idea of selling any of its facilities and insisted that the LCRA should sell power to it for distribution to the various cities.

On February 2, 1939, the LCRA sent a letter to CP&L declaring that it had been advised on numerous occasions of the board's policy to not turn any part of its power over to private utilities, "except such power as may be surplus from time to time over its customers' requirements. In accord with such policy the municipalities have determined to own and operate their distribution systems and to purchase LCRA power." The letter ended flatly with "If you do not agree, we will conclude that you do not desire to avoid competition and do not desire to cooperate with us to avoid duplication of facilities."

The negotiations were destined to drag on for more than a year, but some of the small cities involved were not willing to be that patient. In March, Fredericksburg, Luling, and Waelder announced their intentions to start building their own distribution systems.

Since the negotiations with CP&L seemed to be stalled, the LCRA board voted in March to set aside $650,000 of the PWA's loan allocation for transmission line construction to those three cities. It also stepped up negotiations with the San Antonio Public Service Company to transmit power to all three until the LCRA lines were completed.

In the meantime, Max Starcke was doing one of the things he did best: selling.

15

A Model of Efficiency

Starcke's pride in the product he was selling prompted him to have a plaster of Paris relief map made of the Highland Lakes area, showing the LCRA's dams and its electrification system. The model was twelve feet long, built in three sections, and varied from four to eight inches in thickness. It turned out so well that Mc-Donough wanted to share it with the world — or at least that part of the population which planned to visit the New York World's Fair in 1939.

"You can't imagine how much that thing weighed," said Art Anderson, a civil engineer who learned that truth by taking the map on an adventurous, month-long trip to New York.

Because he had been in the advertising business and was a natural in public relations, Anderson quickly became a man of many hats for the LCRA. And he nearly always wore at least one — literally. He explained:

> My specialty in engineering was steel and I was supposed to be superintendent of steel construction on the Austin Dam. But then they asked me to come to work earlier than I had planned, in the engineering department. I had an office right across the hall from Senator Wirtz and I kept my hat on all the time. One of

143

the boys we hired from The University of Texas finally asked me why I wore my hat in the office. I really had never thought about it. I had just been accustomed to wearing a hat all my life. But I told him that when I went to work there, I didn't know whether I would like the job or not — and I kept my hat on so if I decided to quit, I wouldn't have to delay my departure by going by the hatrack to pick it up . . .

Mister McDonough was from New York, of course, and he and Grover Whalen, who promoted the World's Fair, were big buddies. He wanted to show Whalen and everybody else what we were doing down here so he arranged for this model to be shown at the Fair.

It was so heavy that it took eight men to lift just one section, with its case. We were just finishing up the Austin Dam and the carpenters there built the frame out of two-by-fours. The main piece was four feet square and it sat down in a wooden box, which had a top made of one-inch shiplap lumber. I think it was one hundred and ten screws that held the top down. They had hausers that came out of each corner at first. But then they had to drill some more holes and use some more hausers because four guys couldn't pick it up.

We made quite a display to go with it for the Fair. We had a long painting showing the history of electricity, from Benjamin Franklin and Thomas Edison right on down to the LCRA and our big generators.

McDonough assigned Anderson and Leroy Mundt, an employee who previously had worked for the *Austin American,* to take the exhibit to New York and see that it was set up properly. Mundt also had been an honorary Texas Ranger, a title which in those days entitled him to carry a gun.

Their trip was destined for trouble as soon as McDonough offered them an old flatbed truck from the Austin Dam project. "It had been running over the form nails they'd been pulling and, as it turned out, it had about ten jillion nails in the tires," Anderson remembered. "Of course, we didn't realize that when we left Austin."

They loaded the display onto the truck, covered it with canvas, and put up a sign on the side that said: "LCRA — Going to the World's Fair or Bust!" Also tucked neatly under the canvas was a barrel of dishes belonging to Morgan Potter, a New Yorker whom McDonough had brought to Austin in 1935 as his secretary and top assistant. Said Anderson:

By the spring of 1939, the handwriting was on the wall. It was obvious that Mister Mac and his crew were going to move out of here and Mister Starcke's crew was going to take over. Potter was getting ready to move back to Suffern, New York, and he wanted us to take this barrel of dishes by there on our way to New York City. We really didn't have any choice about that — especially since Potter was handling the purse strings and was the man who handed out all the expense money.

They gave us five hundred dollars in advance. That was supposed to cover the whole trip, including all our meals, hotels and travel expenses. But it took us four days just to get to Little Rock because we had so many flat tires. We went from one filling station to the next one, getting flats fixed. And by that time, I'd paid for so many flats I was broke. I phoned Potter and told him that I needed more money and that if we didn't get some new tires, we wouldn't get to New York until long after the Fair opened. Potter said, "Oh my God, I'll wire you some money."

I've forgotten how much he sent me but I think it was another five hundred dollars. Anyway, we bought some new tires and didn't have any more tire trouble. We delivered the dishes in Suffern and went on over to Queens, where Whalen had his setup.

And all this time, Mundt had his revolver strapped to the steering wheel post on that truck. It was very obvious. We drove into the parking garage in Queens, where we were supposed to report, and there were a lot of people waiting there to apply for jobs at the Fair. They oohed and awed when they saw our sign — and then some more when we got out, wearing boots and khaki pants and big hats. But what really got 'em was when they gathered around the truck and saw that gun. Carrying a gun was a "no-no" in New York. A gun was just something that nobody but a policeman was supposed to carry. And they were all pointing at it.

We asked a man in the office where we could contact the guy we were supposed to see. He said the building was not ready for the display but they would take care of everything. He told us to go over to Manhattan and get a room that had been arranged for us at the Edison Hotel, right on Times Square. He said they'd get in touch with us when they were ready for us.

The hotel clerk looked at us and figured we were a couple of rough old Texas boys and they ought to get us as far away from everybody else as they could. So they put us on the top floor of the Edison. And as it turned out, we were there about a month. By the time we left, the top floor was known as the "Texas Penthouse."

> Leroy put his gun in his suitcase and left it there when we
> went out for supper that first night. When we came back, the
> hotel people had been all through our gear. I guess it was the
> house detective that spotted the gun. So a plain clothesman came
> up there. I never was quite sure whether he was with the Police
> Department or not, but he said he was from the City of New
> York. And he wanted to know if we had a permit for that gun.
> Leroy pulled out his little gold Texas Ranger badge and showed
> it to him. The guy's eyes got big and he backed off, saying, "Yes,
> sir; yes, sir."

Anderson soon found out that it would be at least three or four
weeks before the building which would house the LCRA display
would be completed. He phoned McDonough to give him the news,
and McDonough told him, "Well, Art, you're in my town. Why
don't you just go ahead and have a good time up there until
Whalen gets that darned thing of his set up?" Anderson said they
did exactly that:

> There wasn't a play on Broadway we didn't see. We went to
> all the museums and I even became an airplane pilot while I was
> up there. We were in an art museum and I got in on a drawing
> where I won some flying lessons. I went out to Long Island and
> took them, learning to fly a Cessna.
>
> After the World's Fair, we took that display to the State Fair
> in Dallas. Then, after World War Two, Mister Starcke still
> wanted to use it but it needed to be revised and brought up to
> date. Bob Hammond was building boats out of fiberglass and he
> took us all around his plant, showing us what we could do with
> fiberglass. We brought that model up to date and made three fi-
> berglass castings of it. They hauled those things all over the coun-
> try for fifteen or twenty years.

There was no way to measure, of course, how many World's
Fair visitors may have been sold by that exhibit on the LCRA or on
the public ownership of utilities concept. But at least one person
reaped some benefits from it. During that spring of 1939, the
LCRA faced an uphill selling job on three fronts: it still had to sell
potential customers in rural areas on the benefits of electricity; it
had to sell the Texas legislature on its intention and ability to control floods; and it had to do some strong selling in Washington to
make that flood control possible.

On March 14, 1939, the LCRA board adopted a resolution addressed to the Texas legislature declaring its policy concerning flood control operations with its existing facilities and "especially in connection with the proposal to increase the height of Marshall Ford Dam by adding 78 *[sic]* feet."

The resolution cited the installation of the fifty rain gauges throughout the upper watershed to supplement the thirty-five gauges which had been installed by the U.S. Weather Bureau. But more importantly, it declared that the Board would see to it that Marshall Ford's height was raised, noting that the U.S. Bureau of Reclamation had designed the high dam so that 804,000 acre-feet of capacity in the reservoir could be dedicated exclusively for flood control.

In addition to promising that the proposed 804,000 acre-feet of space would be maintained for flood control, the board reiterated its policy of operating all of its facilities in a manner to provide "the maximum amount of flood control possible, consistent with its financial obligations."

And even without the added height, said the resolution, the LCRA could "control at least eight out of ten floods of the intensity of any of the floods of record originating above its dams."

But on the same day that resolution was adopted, the board recognized that plans for increasing the height of Marshall Ford had hit a major financial snag in Washington. Another resolution was adopted, asking Congressman Johnson to urge that Secretary Ickes request the $17 million needed during the next fiscal year, so that the enlargement work could continue. That resolution noted:

> The Bureau recommended the building of the high dam as necessary to control floods, and Mr. Bashore, one of the Bureau's engineers sent by Secretary Ickes to investigate the 1938 flood, reported that floods of such intensity could not be controlled except by means of the high dam at Marshall Ford . . .
>
> Because of statements of the Bureau of Reclamation and its engineers, the public generally has been convinced that the Authority will have failed to provide adequate flood protection if the high dam is not built . . . The Authority feels it has successfully pioneered a public power policy in Texas, against the attacks of privately-owned utilities, but the enemies of such a program will be furnished a weapon which might enable them to destroy the whole program if the advice of engineers in respect to adequate flood control facilities be not followed . . .
>
> If construction of the high dam is delayed until after the present reservoir has been filled, it will be necessary to drain the

reservoir, resulting in a loss in income to the Authority in excess of $1,000,000.

But legal problems threatened to wash away all hopes for getting the additional $17 million on which the higher dam — and the future of Brown and Root — depended.

The LCRA could not provide the funds because it had used up all of its authorized borrowing power, and all of its potential revenues from the sale of hydroelectric power had been pledged to outstanding bonds.

The U.S. Bureau of Reclamation was prohibited from granting funds for dam construction except for flood control; any money spent to produce power had to be repaid to the government. Since the Bureau already had funded the lower portion of the dam as a flood control project, it could not contend that the increased height would *not* be a power project.

Wirtz came up with the idea of changing the "label" on the 190-foot structure to call it a "power" dam instead of a "flood control" dam. The flood control portion then would become the seventy-eight additional feet still to be built, making the Bureau of Reclamation legally able to pay for it. The trouble with this ingenious solution to the problem was that it ran afoul of all legal precedents and also of a standard formula for determining what portion of a dam was allocated to flood control and what portion to power production.

Lawyers who studied the matter declared that under the formula, only $14,850,000 could be allocated to flood control. The rest of the cost would have to be charged to power production, and that expense would have to be paid back to the federal government. But the LCRA already had agreed to reimburse the government for $9,515,000 and that was the maximum it could expect from power production. That, added to the $14,850,000, totaled $24,365,000. Since the high dam would cost an estimated $27,000,000, that left $2,635,000 which could not legally be paid by either agency.

Fortunately, Congressman Johnson's assets included the friendship of a brilliant young lawyer named Abe Fortas, who eventually would be appointed to the U.S. Supreme Court. Fortas figured out a way to justify everything, using the PWA to fill in that $2,635,000 gap. He convinced Secretary Ickes that the way to handle it was to designate the first 33 feet above the original 190 feet as neither for flood control nor power, but for *both*. The water behind that 33-foot segment could be used for power production until the floodgates were opened, which would be seldom. Most of the time, it could be used to generate power. That way, as Judge Ferguson explained later, the Bureau of Reclamation would pay for the top 45 feet of the 268-foot dam while the PWA paid for the 33 feet just below that.

But while Ickes bought the idea, some congressmen did not. When it came time to tack an essential amendment onto the Rivers and Harbors Bill, exempting the "Colorado Project, Texas" from a clause requiring that projects built by the Bureau of Reclamation be reimbursable, opponents mounted a vociferous attack.

A budding floor fight against the amendment on May 1, 1939, was squelched quickly when Sam Rayburn, then the majority leader and already highly respected, came to its rescue. He addressed the Speaker:

> I do not usually take the floor in connection with appropriations bills because I usually do not know a great deal about them. However, the gentleman from Texas, Judge Mansfield, is unable to be here today. He has written me a letter, a paragraph of which I desire to read, as follows:
>
> "The completion of this dam will prevent the recurrence of disastrous floods which, frequently in the past, have devastated a large portion of my district. I am, therefore, anxious that the House will see its way clear to accept this amendment."
>
> Everyone familiar with the situation in this part of Texas knows that the district represented by Judge Mansfield suffered untold loss last year. If he were here, he could speak better himself than I. But I thought the House would like to know how it affects this grand old man's district and that he is tremendously interested in the House concurring in this amendment.

Rayburn's speech ended the debate and the amendment was approved by the House, 96 to 55. The measure finally was enacted into law on May 10, 1939.

It was, as it turned out, just in the nick of time. Had it been delayed a couple of years more, until the United States entered World War II, the project probably would have been short-circuited for a long time by shortages of critical raw materials.

In 1940, "Mister Sam" Rayburn became Speaker of the U.S. House of Representatives. If they could have, longtime proponents of Colorado River development probably would have elected him king in 1939.

16

The Selling
Circus

The LCRA's strength in Washington was enhanced considerably in December of 1939 when President Roosevelt announced his appointment of Wirtz as undersecretary of interior. At a special meeting on January 2, 1940, the LCRA board granted Wirtz a one-year leave of absence without pay and named Gideon to act as general counsel. The directors made the same type of agreement with Gideon that they had with Wirtz: he would continue as a member of the Powell, Wirtz, Rahut, and Gideon law firm while serving as general counsel.

It was also in December that G. F. Harley, the PWA's resident engineer in Austin, recommended that the board consider phasing out the engineering and construction section. More emphasis should be given to the operating section, he said, since construction of the Austin Dam was nearing completion and the engineering work on the Marshall Ford Dam, power plant, and transmission lines also was winding down.

Harley praised McDonough's work, declaring that he was "due much of the credit for the engineering excellence of LCRA's dams and transmission system."

McDonough, who had agreed originally to stay only through

150

the construction period, told the board on March 4, 1940, that the construction program would be substantially complete by June 30, 1940. He recommended that the construction department be merged with the operations department and that he be relieved of his duties as general manager on May 1.

The board accepted McDonough's resignation "with regret" and voted to combine the two departments. Then, at a later meeting, the directors named Starcke as general manager for the period from May 1, 1940, through December 31, 1942, at a salary of $10,000 per year.

The Austin Dam finally was completed on February 16, 1940, and it began operations on March 11. Dedication ceremonies, with Undersecretary Wirtz as the principal speaker, were held on April 6, 1940 — just one day before the fortieth anniversary of the original dam's destruction by flood.

The Austin Chamber of Commerce had voted in 1937 to name the structure the "Tom Miller Dam" in honor of the Austin mayor who worked so long on the reconstruction project. When the problem of naming a federally financed project for a living person arose again, a plaque honoring Miller was unveiled but the name remained unofficial.

Wirtz cited the new dam as an impressive symbol of river development.

> This visible demonstration of what must be done to preserve our natural resources through their use cannot fail to encourage us in the accomplishment of other conservation work that we have before us. We see here a development that does not use up and does not exploit but that uses and preserves . . . And all the farmers and townspeople who are using the cheap electricity provided by these great dams, and who have received a measure of flood protection from their construction, have benefitted themselves and their children.

That spring of 1940 turned out to be an eventful one for the LCRA.

Congressman Johnson, disappointed by the failure of his plan to have only one coop — and that operated by the LCRA — covering his district, kept insisting that the coops would be more efficient and their electricity more economical if their facilities were combined and operated by the LCRA. He repeatedly urged the boards of the Pedernales Electric Cooperatives, Inc., and the Lower Colorado River Electric Cooperative, Inc. (LCREC), to work out such an agreement with the LCRA.

The LCRA had acquired a number of rural distribution lines as well as systems in many smaller rural towns when it purchased the Texas Power and Light Company's facilities in sixteen counties. Costly duplications loomed as both the PEC and the LCREC began constructing lines to reach their members and to connect their systems with the LCRA.

On March 4, 1940, an agreement was reached for the LCRA to sell to the two coops the rural transmission and distribution lines it had acquired from TP&L. This was subject to the approval of both the RFC and the REA, which was obtained within a short time.

Lengthy negotiations were climaxed by the signing on May 31, 1940, of a contract calling for the PEC and LCREC to turn over all their property to the LCRA to be operated by the LCRA. The LCRA agreed not only to perform all necessary construction work but even to keep the accounts on each of the customers, send monthly bills, and collect the payments.

Congressman Johnson praised the agreement in a press release, predicting that the resultant economies would reduce the monthly minimum bills from $2.50 and $2.45 to $1.25. And this, he said, would make electricity available to many farm families which could not afford it before.

Another set of marathon negotiations also ended that spring, when the Central Power and Light Company finally agreed to sell to the LCRA its properties in nine cities and adjoining rural communities. The price of $1,385,926.47 was set by Francis P. Quigley, a PWA engineer sent from Washington by Ickes to make the appraisal.

The agreement came only after the LCRA board, faced with the need for developing income from the sale of electric power which was now available from its hydroelectric plants, issued an ultimatum. The directors notified the CP&L that "unless a definite agreement shall have been reached within 45 days from May 6, 1940, LCRA will proceed with construction of its own high lines and will notify the cities of the affected area so they may proceed with the construction of their own distribution facilities."

Just before the deadline, on June 17, the CP&L agreed to sell the properties at the price set by Quigley.

Meanwhile, Max Starcke was exploiting fully the "electrical circuses" staged by the Rural Electrification Administration. Large circus tents were set up for three- or four-day stands in various towns to house electrical appliance demonstrations by major manufacturers. Depression-battered farmers still were skeptical about investing even so much as $5 a month in electricity. But while their wives stood at one end of a circus tent marveling at washing machines and cooking ranges and hot water heaters, many of the farmers were at the other end watching demonstrations of power-driven pumps and milking machines. Anderson recalled,

> In many cases it really was a "hard-sell" situation. Even a cooking range back in those days cost about two hundred and fifty dollars. And a lot of these farmers were just barely getting by. To go out there and tell an old boy, "You've got to spend some money to make money," was a hard thing to do.
>
> I remember we had one house between Austin and Bastrop that was not more than a hundred feet off the highway. The coop line came right by there but the guy who lived there wouldn't hook on. We did just about everything we could. We would have given it to him if we'd had to because it was so obvious that he was right on the highway and didn't have electricity.
>
> He was an old Dutchman and we went to him and tried to talk him into signing up. He just sat there and shook his head and said, "I live all my life here. My Daddy live all his life here. We don't have electricity. We don't starve. We do all right."
>
> "But think of your wife," we told him. "Look at her out there on that washboard." He looked at her and said, "What's the matter with her?"
>
> We used to tell a joke about a Cajun crayfisherman over in Louisiana. They finally talked him into hooking up but for months he only had a minimum bill — about two dollars and fifty cents a month. They went out to see if he was jumping the meter or something. They confronted him and asked why his bill wasn't bigger. "Don't you use the electricity?" they asked. "Yes," he said. "I use it every day. When I wake up, I turn on the light, find my lantern and then turn it off."
>
> This pretty much represented the feelings of a lot of people. Many of them thought that about all electricity was good for was to provide light to read by. And light for reading proved to be our biggest selling point. But once you had the house wired, it was kind of a race between the radio and the refrigerator as to which they got first. Usually, the wife won with the refrigerator. Of

course, if you had a dairy or a place where a lot of food would spoil, a freezer came along pretty fast.

With the aid of REA financing, most of the people in rural areas did not have to pay any hookup costs; their only expense was the charge for the electricity they used. But it still was difficult to convince many farmers that using electricity actually could increase their profits by performing a great many chores faster, cheaper, and more efficiently than they could be done by hand.

But the LCRA was so proud of its efforts and success in that field that it gave itself a statewide pat on the back with a press release issued on December 13, 1940, which said:

> Behind rows of neatly aligned figures in a bookkeeper's ledger at the Lower Colorado River Authority's offices in Austin these days lies a graphic story.
>
> It's a story of a once unruly, always threatening river now a docile servant of thousands of Texas citizens; of water that once coursed madly and devastatingly to the sea now making electricity to light homes, to cook meals, to refrigerate food, to carry running water to a housewife's kitchen, to make possible the hundreds of conveniences of modern living in thousands of homes where once the cost of these things was prohibitive.
>
> These columns in the bookkeeper's ledger set down in cold, unromantic figures the story of how rate reductions placed into effect by the LCRA have brought a 32.75 per cent increase in residential use of electricity in eleven towns served directly by the Authority with low-cost power from its flood control, water conservation and power generation project on the Colorado River above Austin.
>
> Service in these towns — Bastrop, Bellville, Brenham, Burnet, Giddings, Kerrville, Lampasas, Llano, Lometa, San Saba and San Marcos — was started by the Authority in September 1939.
>
> They were a part of electrical distribution purchased from the Texas Power and Light Company in a 16-county area accessible to the LCRA project, and are those that have not been sold to municipalities.
>
> Shortly after it took over operation, the LCRA placed into effect a 37 percent residential rate reduction and a 20 percent commercial rate reduction.
>
> That started, almost immediately, an ever-increasing use by householders of this low-cost electricity generated at their own dams by their own river.

Comparing its consumption figures for the first 11 months of 1940 with the first 11 months of 1938 the LCRA shows an average increase in residential consumption of 1,592,301 kilowatt hours or 32.75 percent increase . . .

Figures for 1938 cover the last full year of operation in these cities by the private power company and were the only ones available to the LCRA upon which they can make their comparisons.

Along with making more extensive use of electricity possible by those who already enjoyed service, the low rates, too, have made possible in these towns use of electricity by many families who previously had not been able to take service. Increase in customers during the 1940 period over the 1938 period was 10.77 percent . . .

While the figures used here, in themselves, tell a graphic story they are but a part of all the story of the benefits brought to Texans by the harnessing of the once unruly Colorado.

There, too, is the story of prevention of floods that for years took an average annual toll of approximately $4,000,000 in property, many human lives, and tons upon tons of essential top soil; of water that once made its way hurriedly to the sea which now is conserved for use in the times when floods are replaced by drouths; of great lakes creating the variety of recreational activities water makes possible.

In still more books is written the story of thousands of rural dwellers, to whom electricity once was too inaccessible and too expensive but to whom now low-cost power from the Lower Colorado River Authority, flowing through the outstretched arms of REA lines, brings the comforts and conveniences of their city cousins and new avenues for making profits out of their farms and ranches.

Salesmanship efforts at the grassroots level picked up steam in 1941 as even more power became available. The LCRA began generating electricity at Marshall Ford on January 27, 1941. On February 21, the LCRA board voted unanimously to name the structure "Mansfield Dam" in honor of Congressman Mansfield, who had just celebrated his eightieth birthday.

The huge dam, on which construction had begun December 3, 1936, finally was completed in May of 1942. Equivalent in height to a twenty-five-story building, it is one of the largest masonry structures in the world. It is more than a mile long, impounding a maximum of two million acre-feet of water and creating a lake 65 miles long and 8.5 miles wide in its maximum reaches. Its foundation width is 228 feet and its average thickness 125 feet. More than three miles of inspection tunnels — some five feet by seven feet and others six and one-half by eleven — lace its interior. There are 24 eight-and-one-half-foot openings with 48 flood gates. Of the total cost — $29,000,999 — the Bureau of Reclamation paid $23,294,055 and the LCRA only $4,914,163.

By the time those cost figures came out, the LCRA public relations representatives had some much smaller numbers which they preferred to publicize. Early in 1940, they had decided to keep detailed records on a single farming operation to use as an example of what electricity could do not only to promote comfort and efficiency but to increase profits as well. Art Anderson recalled their picking a dairy farm owned by the R. L. Sansom family about two and a half miles west of Buda, in Hays County.

> We kept such close records on them that we practically lived with them. I think Mister Sansom was spending something like thirty dollars a month just to chill his milk and get it picked up. He had to get up at three o'clock in the morning to milk his cows, then he had to chill the milk and take it a mile down the road so the milk company could pick it up there in refrigerated trucks and take it into town. That was before he got electricity, of course . . .

The *Austin Statesman,* in an article published on June 9, 1941, cited the Sansom case as an outstanding example of low-cost electricity's benefits. It quoted Mrs. Sansom at length on the benefits of her electric range and refrigerator and added:

> In the Sansom home, too, are electric lights that eliminate the eyestrain caused by the old type of kerosene lamp. There are the various small appliances such as fans, a radio, iron and waffle irons.
>
> An electric water heater in the winter eliminates the inconvenience of having to heat water for household uses by the old-fashioned kettle and pot method. Instead, her electric water heater makes possible gallons of hot water always available for the mere turn of a faucet.
>
> Not only have the Sansoms put electricity to use in the home.
>
> Mr. Sansom is finding it saves time and money for him in his farming operations. In the place of a gasoline powered motor now is an electric one that goes into action with no more trouble than turning of a switch.
>
> An electric milking machine has replaced the former system

of milking by hand, while in the place of an ice cooling system is an electric one. It requires no filling to provide refrigeration, electricity does it all.

To say the Sansoms are sold on low-cost electricity would be understating the case. They're enthusiastic boosters of it.

And by figures they can show it pays. Where gasoline and ice for operation of the dairy once cost them $30 a month, they now pay an electrical bill of about $9 a month.

And in this electrical bill, too, is not only the dairying work that formerly cost $30 for gasoline and ice but also lighting, cooking, refrigeration and other conveniences for the home.

So for far more than they formerly got, the Sansoms pay far less money.

In fact, the difference between the cost of electricity and the former cost of gasoline and ice alone pays the cost of the appliances.

The LCRA had demonstration pictures, along with detailed accounting of the Sansoms' expenses, to make a persuasive sales pitch — one that was used over and over, to great advantage.

With all the selling that was going on externally, Starcke still sold himself, apparently without any conscious effort on his part, to his employees. Or most of them anyway. Anderson recalled:

He had an open door policy to everyone, and there was pretty much a family atmosphere around the LCRA as long as he was there. And when he was general manager, there was not a single person hired by the LCRA — until after World War Two, at least — that had not been interviewed personally by him. You could tell him you had a man you wanted to hire and he would say, "Send him on up." He'd talk to the guy and decide if he was "our kind of people." And he was a good judge of people.

There was nothing that went on at the LCRA that Mister Starcke didn't tell everybody about. If the board met in the morning, by coffee break time that afternoon all the employees knew everything that had gone on. Starcke would hedge-hop from one table to another during coffee break and say, "Well, I think we got this done . . ."

Whenever he went on a trip, he always sent back postcards to the employees telling them all about it. We were all part of his family and he shared everything with us. We all called him "Mister Starcke" — because we respected him so much.

When I came home after World War Two, Mister Starcke told me that the most heart-breaking thing that ever happened to him was when this union leader came into his office one day, started pounding on his desk and telling him the LCRA had to pay more money. Mister Starcke told him he thought those people worked for the LCRA. And the guy said, "They work for me now."

Mister Starcke said he didn't know anyone was unhappy. But he called a special meeting of the board. He found out this union guy had gone to all the different departments and talked to one or two disgruntled people in each to get his foot in the door. Mister Starcke couldn't understand why they hadn't just come to him if they were unhappy.

Anyway, this guy said they were going to shut the place down. . . .

I was on Iwo Jima at that time, and we heard on the island radio that Mister Starcke had fired everyone who went on strike. There were a lot of labor union people in my outfit over there but they were disgusted with the strikes that began breaking out back home as soon as the war ended. A lot of them knew I had worked for the LCRA and when they heard the news that those strikers had been fired, I became a local hero.

17

On the
Warpath

Early in 1940, the Austin Chamber of Commerce's Military Affairs Committee began exploring the possibility of getting a military base built in Central Texas. The committee finally decided that the best available spot was an area of about 50,000 acres in the Bastrop-Elgin-Smithville vicinity.

That fall, the committee sent a four-man delegation, headed by Max Starcke, to present the idea to Gen. Hubert J. Brees, commanding general of the army's 8th Corps Area with headquarters at Fort Sam Houston in San Antonio. Starcke was accompanied by Walter E. Long, A. B. Spires, and John H. Frederick.

Long later reported that the group received an unusually warm reception when General Brees learned that it was proposing a base not immediately adjacent to Austin but thirty-six miles away. He said he was tired of listening to so many similar groups which insisted a base must be built in or adjoining their own cities.

The army announced on January 2, 1941, that it planned to establish a camp at the Bastrop site. A contract for the necessary surveys was signed on July 29, 1941. Construction contracts totaling $25 million for a camp to house between 30,000 and 40,000 troops were let on January 19, 1942. With World War II under

159

way, approximately 18,000 employees were hired to construct "Camp Swift" on a 55,906-acre tract in not more than 108 working days.

The LCRA was to furnish electricity for the camp. With a frantic, almost around-the-clock effort, the necessary transmission lines and other facilities were installed — and the Authority's emphasis shifted suddenly from selling power to supplying it.

In late January, the LCRA's ever-eager publicity department issued a press release declaring:

> An example of how Americans will get their Herculean task of defense production done this year has been set by workmen for the Lower Colorado River Authority.
>
> The LCRA, whose low-cost electricity produced at its dams on the Colorado River above Austin will play an important part in defense works, turned out a two-weeks defense job in five days.
>
> Called upon to supply power for the gigantic army camp to be built in Bastrop County, linemen set to work in sub-freezing weather with an all-out will . . .
>
> Typical of the spirit of men was that of F. W. Hill, lineman. He refused to leave the top of a pole where he was installing a small transformer until he completed the job.
>
> "I'll get a lot of warm satisfaction out of knowing this job is done," was Hill's reply to the foreman's urgings to come down and warm himself at a camp fire.
>
> With the difficult job of installation of lines, transformers and other electrical equipment to bring power from the Marshall Ford Dam into the army camp completed, the Authority already had begun serving the quartermaster's quarters erected on the site and was ready to supply electricity for the construction crews that will move onto the site shortly.
>
> Plans, meanwhile, were going forward at the Authority's headquarters here to bring service to the large magnesium plant near Austin, which will soon be built as a part of the war effort to produce the vital magnesium metal for the nation's huge airplane production program.
>
> "President Roosevelt said in his address on the State of the Nation that this country would have to work 24 hours a day and seven days a week to get this war production job done," William B. Arnold, chairman of the LCRA Board, declared in commending the line crews for their work on the Bastrop army camp job.
>
> "We're proud of the spirit our workmen showed in pitching into that Bastrop job and accomplishing just that kind of produc-

tion. It didn't take any pep-talking or any whip-cracking. They just turned in and did it and that's the kind of action all over this country by working men that's going to show Hitler and his gang Mr. Roosevelt knew what he was talking about when he told them the number of planes and ships and guns we would build this year and next year to beat them with."

By the time World War II ended, the LCRA would not be quite so proud of all its electrical workers. But the war effort had brought people together in a remarkable demonstration of unity and determination.

Lt. Col. L. A. Kurtz, Camp Swift's first commanding officer, arrived from Fort Sill, Oklahoma, with five other officers on April 23, 1942 and formally assumed command on May 3. Col. Oscar Parke Houston became the commanding officer a short time later and eventually joined Long in writing an unpublished history of "Camp Swift, Texas." They noted that the huge base was built in near record time and ultimately housed 44,000 troops at its operational peak.

About the same time Camp Swift was activated, an ambitious young high school student named Elof Soderberg heard that the LCRA had an opening for a mail boy. Soderberg was working part-time as a copy boy for the *American-Statesman,* which had a building on the northwest corner on 7th and Colorado streets. He had gone to work there during the summer of 1941 for $3 a week. After a few weeks he told the editor, Charles E. Green, that he felt he was underpaid, especially in view of the long hours he worked, and deserved a raise.

"Well," said Green, "since you were brave enough to come in here and talk to me about it, I'll raise your pay to $5 a week."

But when Raymond Brooks, an LCRA board member who also was a political and editorial writer for the newspaper, told him about the mail job opening, Soderberg went almost immediately to the LCRA offices on the third floor of the *American-Statesman* Building. He was interviewed by Beverley Randolph, who hired him for the magnificent sum of twenty-five cents an hour.

"That was a tremendous increase from the five dollars a week I was making at the paper," recalls Soderberg, who rose through the ranks to chief engineer and finally general manager of the LCRA.

Soderberg quickly became more than just a mail boy. With

airline passenger service to Austin almost non-existent, he frequently drove Starcke to Houston or Dallas to catch a plane, or met him at one of those airports to drive him home. And Starcke, he recalls, always was in a hurry on those trips, most of which were to New York and Washington.

"I remember one time when we were going to Houston," said Soderberg, "and I was trying to stay under the wartime speed limit of thirty-five miles an hour. Mr. Starcke told me to 'step on it — and don't worry about a ticket; if you get one, I'll pay it!' "

The two became close enough that the brash youngster once asked Starcke if he would mind his sitting in the general manager's chair for a moment.

"Of course not," said Starcke, "but why do you want to?"

"I just want to see how it feels," Soderberg replied, "because I intend for it to be mine some day!"

Soderberg graduated from Austin High School in January of 1944 and went into the Army Air Corps as an aviation cadet two months later. After the war, he entered the University of Texas Engineering School and went to work part-time, mostly testing electrical meters, for the LCRA in 1947. He was on the bottom rung of the ladder to the general manager's chair. By the time he reached the top, he would not be sure that he still wanted it.

By 1943, the LCRA's power sales were generating enough revenue to warrant refinancing the $21,635,000 in bonds held by the federal Reconstruction Finance Corporation. The Authority went into the private market and accepted a low bid submitted by Stranahan, Harris and Co., Inc., of New York on $21,635,000 in bonds at interest rates low enough to save $4,034,411.75 through the refunding arrangements.

Paying off the old bonds with money from the new ones also eliminated the LCRA's indebtedness to the federal government. But the LCRA's financial dealings also drew fire, in the form of a legislative investigation, during 1943. That probe was aimed at a possible loss of tax revenue resulting from a complicated deal involving the City of San Antonio, the Guadalupe-Blanco River Authority, and the LCRA. The City of San Antonio sold its Comal generating plant to the Guadalupe-Blanco Authority, which then leased it to the LCRA.

Perhaps the main result of the investigation was some valuable experience in dealing with the legislature for Sim Gideon. He was expected to handle the situation by himself, and he wasn't quite sure who was behind it all. "There was some feeling, politically, that it might have been aimed at embarrassing Wirtz and some of those who were supporting Roosevelt for a fourth term," he recalled. "But that's just conjecture . . ."

Regardless of the initial objective, the investigating committee's majority report proposed several minor amendments to the LCRA Act; all of them were acceptable to the LCRA board and were approved by the legislature. In connection with the possible loss of tax revenue, the report declared:

> On the other hand, the rates of the Lower Colorado River Authority are much lower than rates charged by private utilities. While the State, counties and municipalities are deprived of much revenue they would receive from private utilities, these losses are more than compensated by the greatly reduced rates for light and power enjoyed under the public utility system.

Looking back forty years later, Gideon commented: "Maybe that deal *was* illegal, since the LCRA at that time, under the law, did not have the right to *own* a steam plant."

That lack of authority to own steam plants ultimately became one of Gideon's greatest challenges. The private utility companies wanted to keep the LCRA restricted to generating hydroelectric power — which now accounts for only about twelve percent of the electricity it produces.

But during World War II, the thought of such problems probably did not even occur to Gideon. Throughout most of the war, a long-festering sore spot involving the International Brotherhood of Electrical Workers lay dormant. During the early days, the LCRA's labor relations were subject to a standard clause applied to both private and public agencies working on PWA projects. It declared:

> Employees shall have the right to organize and bargain collectively through representatives of their own choosing, and shall be free from the interference, restraint or coercion of employers of labor, or their agents, in the designation of such representatives or in self-organization or in other concerted activities for the purpose of collective bargaining or other mutual aid or protection. No employee and no one seeking employment shall be required as a condition of employment to join any company union or to re-

frain from joining, organizing, or assisting a labor organization of his own choosing.

After the LCRA's main construction program ended in 1940, the Authority was no longer subject to the PWA regulations. Local 520 of the IBEW began pressing for extension and expansion of the work agreement McDonough had signed on June 13, 1938, which covered construction employees.

On August 9, 1940, the LCRA asked Attorney General Gerald Mann for a ruling on the legality of its signing a contract proposed by IBEW-520 on behalf of the electrical workers. On September 9, the attorney general ruled that the board could not enter into such a contract because it "would abridge the exclusive power of the Board to perform the official duties thus expressly entrusted to them."

Among the supporting cases cited in Mann's opinion was *Bowers vs. City of Taylor* (Texas Commission on Appeals), 15 S.W. 2nd 520, which declared: "It is well settled that no governmental agency can, by contract or otherwise, suspend or surrender its functions, nor can it legally enter into any contract which will embarrass or control its legislative powers and duties, or which amount to abdication thereof."

The opinion also noted that the National Labor Relations Act did not apply to national, state, or municipal employees.

Some members of the LCRA board still felt that it should issue a memorandum concerning its policy on labor, particularly in regard to the rights of employees to organize and bargain collectively. Thus, on December 9, 1940, the board adopted a resolution reiterating its labor policies, as follows:

> 1. It shall be at all times the policy of this body to pay prevailing wages.
> 2. That the right of employees to organize and bargain collectively through representatives of their own choosing, as established by federal regulation, is recognized by this body and all employees of the Authority are assured of this right.
> 3. That the management of the Authority is hereby instructed by the Board of Directors to notify all employees that this right of collective bargaining is recognized.

That policy also was submitted to the attorney general, and he ruled on December 18, 1940, that it was legal since "no attempt is made to bind the Board, irrevocably, to a state course of action calculated to destroy or to impair the free exercise of judgment and discretion of the Board in respect to the performance of its duties."

On April 4, 1941, IBEW-520 asked that it be recognized as the exclusive bargaining agent for the LCRA electrical workers. In a letter to the board, the union said a large majority of the electrical workers were members of Local 520 and had selected that union as their exclusive bargaining agent.

The board voted, 6–3, to reject the proposed contract on grounds it was prohibited by the attorney general's ruling of September 9, 1940.

The IBEW took its case to the U.S. secretary of labor, which referred it to the National War Labor Relations Board (NWLRB). But on December 23, 1942, the NWLRB ruled in a similar case involving the City of Newark, New Jersey, that "as a matter of law, it [the NWLRB] does not have jurisdiction over labor disputes between state governments, including political subdivisions threof, and their public employees."

That put the matter to rest until August of 1945. Shortly after World War II ended, but while the nation still was on a war footing, Local 520 again began to press for recognition. Harry Bernhard, its business manager, sought a meeting with the board to discuss collective bargaining but was told the board would stick to procedures outlined in a detailed "Working Memorandum" it had adopted in 1941. That called for the board's Labor Relations Committee to meet with employees on all questions or differences that might arise.

Bernhard and a union committee composed of Roy Atwood, C. M. Cranford, Alva Woodward, and L. Buck Baker then drove to Bay City to see P. R. Hamill, chairman of the board's Labor Relations Committee. Hamill, a banker who had been a member of the board since July 1943, had just become chairman of the committee. He agreed initially to call a meeting of the panel for September 5. But after conferring with Starcke and other board members, he wrote the union committee on August 20, 1945, that the September 5 date would not be convenient. And he recommended that the union members write to the Labor Relations Committee as LCRA employees, not as a committee representing the IBEW.

That really rubbed the union the wrong way and it quickly rejected the idea. On September 7, Local 520 instructed the union committee to call a strike if it could not obtain a meeting with the LCRA board as union members. During the next few days, it asked the 130 LCRA workers who belonged to the Union to sign pledges ratifying the strike action; 99 did, according to IBEW officials.

By late September, Governor Coke Stevenson became so concerned over the labor strife within the LCRA that he asked State Labor Commissioner Leonard Carlton to call a mediation conference. According to Carlton, the union expressed a willingness to go along with such a meeting but LCRA officials told him they felt there was nothing to mediate.

The *State Observer* reported later, on October 22, 1945, that Stevenson had warned Starcke of the strike possibility but was told that he "would manage somehow."

The strike came at 12:15 A.M. on Monday, October 1, 1945. Electrical workers pulled the switches in LCRA power plants and cut off electricity for approximately 250,000 people, including many in hospitals and military installations. The City of Austin suffered only minimal effects, since it had its own steam generating plant, but its water supply was jeopardized.

The *Austin Statesman* reported the situation in its October 1, 1945, edition:

> A strike of union employes of the Lower Colorado River Authority Monday at 12:15 A.M. closed down hydroelectric generating plants operated by the Authority on the Colorado River at Austin, Mansfield, Inks and Buchanan Dams.
>
> Recognition of their union was the declared objective of the strikers, who claimed the authority would not recognize their right to organize and bargain collectively.
>
> Harry Bernhard, business manager of Local 520, International Brotherhood of Electrical Workers (AFL), said between 130 and 140 members of the local had left their jobs at the four plants. Picketing on an orderly basis was started immediately at Austin Dam.
>
> Max Starcke, general manager of the LCRA, said the strike was called without notice. He said that the meeting of the bargaining agency of the operative employes had been requested but not granted by LCRA.
>
> The LCRA employes had not asked for any wage increase, Starcke said, nor had they complained of working conditions, he added.
>
> "This strike was called without notice to us," he asserted.

"The employes have been asking recognition of the union as their exclusive bargaining agent. We have not granted that."

Electric service was cut off from the army's regional hospital and all of Camp Swift, the veterans hospital at Kerrville, Bergstrom Field, the San Marcos army air field, the rural area around Austin and from 23 cities and towns served by the Authority, and about 25,000 rural customers and residents of about 50 small communities served by rural electric cooperatives.

Approximately 10 private hospitals also were without power.

Recently, the IBEW (AFL) and a group of authority employes presented a demand that the IBEW be recognized as the exclusive bargaining agency of plant and electrical workers. This was rejected by the authority with the statement that as a public agency it could not delegate hiring and firing or other official responsibility to any agency. This statement was made by all the directors of the authority, except two who are in the armed services and were absent.

No grievance, pay question or other request of employes has been made, Mgr. Starcke said.

In at least four places, the Austin plant, Peters sub-station, San Marcos and Lampasas, when the switches were closed, it was found that power was not resumed, due to other things than the opening of the switches alone. These other causes of interruption were being investigated . . .

It turned out that many of the transmission lines and other equipment had been grounded, making it impossible to restore power by simply throwing the essential switches. An investigation of apparent sabotage began immediately.

The most severe damage from the power outage was inflicted upon the hospitals, locker plants, dairies, hatcheries and water pumping plants. But the inconveniences suffered by so many people produced a wave of criticism directed at the IBEW.

The LCRA board held an emergency meeting that Monday afternoon and took a strong stand. Members adopted a resolution expressing their feelings and intentions:

That the Authority continue to operate its properties and facilities according to the principles heretofore declared and adhered to by the Authority;

That it continue the operation of its plants and facilities with employees remaining in its employ and such other qualified employees as may be necessary;

That employees who walked off their jobs this date, failing

and refusing to perform their duties for which they were employed thereby severed their employment and will not be returned to such former employment;

That the Authority requests the cooperation of peace officers of the state, the counties and the cities in protecting the property of the Authority and the continued and uninterrupted operation of this public state project;

That the interruption of service caused by various acts of vandalism during the morning hours of Monday and throughout the day is being remedied with utmost dispatch, and that with adequate protection of the properties against further acts of vandalism, service will be restored in full at the earliest possible moment.

The vote was 7–0, with board members Hamill and John B. Connally, the future governor, absent; they were still serving in the armed forces.

During that fateful Monday, employees and supervisors who did not go on strike were able to restore service to a part of the area that had been blacked out. After thirty hours, power was flowing again to approximately ninety percent of the service area. Lampasas, the last city to get its lights back on, began receiving electricity again at 11:00 A.M. on Wednesday — fifty-nine hours after the power had been cut off.

The board's declaration that workers who did not report on their next regular shift would be terminated stuck. About ninety failed to report — and they were fired. That met with the approval of Governor Stevenson, who said at a news conference on October 2: "Any man who commits sabotage on the property on which he is employed should not be re-employed."

While most of the public reaction to the strike was highly critical of the strikers, the feeling was not unanimous. Even Mayor Tom Miller, whose efforts to mediate the dispute were turned down by the LCRA, criticized the manner in which the board was handling the matter.

The *State Observer,* an ultra-liberal publication, devoted an entire issue to the case for the IBEW but the union received very little public support for calling a strike against a public utility without warning. The suspected acts of sabotage, which delayed restoration of power, did not help its case with the public.

The LCRA quickly began replacing the discharged strikers. About twenty new workers were hired each month until the ninety who had been fired were all replaced. Quite a few of those hired were former employees who had entered the armed services and returned home; they were delighted to find jobs. Other war veterans who had gained electrical experience in the armed services were hired and, within six months, the working force at the LCRA was back to normal strength.

But the most enduring, important aspect of the strike came in 1947, when the 50th Texas legislature passed a bill making it illegal for public officials to grant collective bargaining contracts or to recognize union representatives as bargaining agents. The bill also outlawed strikes by all public employees. It was passed by the House, 107 to 19, by the Senate on a voice vote, and was approved by Governor Beauford H. Jester on April 19, 1947.

Another measure, Senate Bill 178, declared that willful stoppage or interruption of public utilities furnishing electricity, natural or artificial gas, or water to the public was a "public calamity which cannot be endured." Picketing and threats "with the intent to disrupt service" were forbidden and district judges were directed to issue injunctions in cases of such picketing. Willful damage or sabotage of public utility equipment was made a felony punishable by two to five years in the penitentiary. A conspiracy of two or more to violate the act also was made a felony.

The governor was directed to "exercise all the power available under the Constitution and the laws of the State to protect the public" against interruptions of public utility service through strikes. But the act also stated that workers still had the right to quit work or refuse to report to work.

That bill passed the House by a vote of 117–2 and then was approved in the Senate on a voice vote. Governor Jester signed it into law on April 14, 1947 — climaxing what amounted to the most turbulent internal combustion ever experienced by the LCRA.

18

Missions of Mercy

Sim Gideon had nothing but praise for the job Starcke did in selling electricity. But there was a time when he thought perhaps the demand created by the combined efforts of Starcke and World War II might be more than the LCRA ever would be able to handle. Gideon said of Starcke:

> He was a very likable fellow, and he did a very good job that needed to be done in creating a market for the power. They thought — and that included Congressman Buchanan, in his letter to the legislature urging them to pass the LCRA bill — that these dams would create enough power to take care of this whole area and send some up to Dallas and Fort Worth.
>
> Nobody dreamed you'd have the power demand you had after World War Two. They had the factories running and they had to keep them going — so instead of building tanks and guns and airplanes, they built refrigerators and washers and dryers and dishwashers and things like that . . . And that created a tremendous demand for electricity.

The new way of life created by electricity contrasted sharply with Gideon's boyhood in Coleman. Only a drop light had hung in

170

the center of each room, and there was no refrigerator or any other appliance for a family of eight children.

"After the war, air conditioning came along and we soon got to the point where Austin — even before it started growing so fast — was using more electricity than the entire state had used before the war," said Gideon.

The LCRA began showing a profit in 1946, a fact which did not escape the attention of Congressman Johnson. In September of 1947, he noted that the LCRA had shown a two-year surplus of $1,297,223 from sales of hydroelectric power and asked the board of directors what it intended to do with the money. As usual, he had a suggestion.

Johnson reminded the board that it had overlooked, up until that time, a clause in the original LCRA Act authorizing the agency to "forest and reforest" the watershed area and also to aid in the prevention of soil erosion. Since the LCRA was a non-profit organization, he said, the profits must be used either to reduce rates charged for electricity or to implement soil conservation measures. And he had detailed plans for the latter.

He urged the board to hire the best man it could find to serve as "director of conservation" and to set up experimental conservation projects on LCRA lands. Then, he said, the board should buy heavy earth-moving machinery for terracing and other soil work. That machinery could be rented at cost to soil conservation districts within the LCRA area, or to county commissioners' courts where there were no conservation districts.

An article by Gordon K. Shearer of the *United Press*, then a widely circulated wire service, reported on September 25, 1947:

> AUSTIN, Tex., Sept. 25 (U.P.) — Texas river authorities bid fair to become the big dog in soil conservation in Texas.
>
> Tucked away as paragraph "d" of a section of the Act to Create the Lower Colorado River Authority is power "to forest and reforest and to prevent and aid in the prevention of soil erosion."
>
> The LCRA act was the model act from which legislation for most of the river authorities is copied.
>
> In the long fight in the Texas Legislature when the LCRA came into being, this phase of its powers was entirely overlooked until Rep. Lyndon B. Johnson, in whose congressional district most of the LCRA area lies, called it to attention of the directors . . .

After outlining the details of Johnson's plan, Shearer's article went on to say:

> To get it down to the individual farmer the equipment will be assigned to work on his land by the district conservation agency or the commissioners court. With it will go a trained operator. The farmer will pay the cost of the operator's salary and fuel cost — about $2 an hour. And best part of it is, that under a federal act the farmer can get the $2 back from the U.S. Department of Agriculture as a reward for his soil conservation activities.
>
> The plan already is being used along the San Jacinto River and is expected to spread to other river valleys of the state . . .

Johnson also recommended several other steps, including consultations by board members with local officials in the ten counties to determine what else should be done, such as research on new uses for cedar in the Hill Country and methods for eliminating weeds on rice farms. He suggested that check dams be built on several Colorado tributaries and that a comprehensive program of resodding be established.

His proposals were adopted on September 17, 1947, by the LCRA board. Caesar "Dutch" Hohn, a prominent Texas agricultural leader who was then with the Texas A&M Extension Service, was hired as conservation director. Hohn, a football star at Texas A&M from 1909 until 1911, had been with the Extension Service since 1920 and was state supervisor of its farm labor program.

The soil conservation program was launched formally on November 2, 1947, at Buchanan Dam; Johnson, Hohn, Starcke, and a host of other speakers on the program promised all-out efforts to make it successful. The meeting, climaxed by a barbecue which Johnson hosted, was described in an article written by Glenn M. Green, Jr., in the *Austin American* of November 3, 1947:

> BUCHANAN DAM, Nov. 2 — State, national and county soil conservation officials here Sunday pledged cooperation to the first locally-sponsored conservation program in the nation as the LCRA-GBRA 24-county reclamation project was officially launched by Congressman Lyndon B. Johnson.
>
> The LCRA has voted $100,000 and the Guadalupe-Blanco River Authority $50,000 to pay for the big job.
>
> One hundred and fifty-four county officials, rural electric co-op directors, officers and board members of the LCRA and

GBRA met at Buchanan Dam to give the program a send-off. They heard outstanding conservation authorities from over the state who pledged their cooperation and commended Johnson for originating the plan.

"It is time to stop conversation and start conservation," Johnson said, but he warned that the program would take time.

"We didn't put lights in 18,000 rural homes overnight. It took nine years," he said.

"In conservation, we have reached the point where we have the money in the bank, the directors and plenty of land to work with."

Pointing to the need for conservation, Johnson said that the United States, with less than six percent of the world's population, possesses 50 percent of the world's goods and is being called on to increase its production further.

Outlining his four-point rural program, Johnson said he has worked for and would continue to work for government-guaranteed prices, rural electrification, all-weather rural roads and soil conservation . . .

Less than a year later, Johnson won his still-controversial election to the U.S. Senate by that famous margin of 87 votes, out of 988,295 cast, over Governor Coke Stevenson. Johnson trailed Stevenson in the first Democratic primary, held on July 24, 1948, by a margin of 477,077 to 405,617. On September 1, the Texas Election Bureau said complete, unofficial returns from the August 28 runoff showed Stevenson had won by 113 votes out of nearly a million cast. But two days later, Jim Wells County filed an amended return, claiming that a re-canvass showed 203 votes cast in Box 13 in Alice (Jim Wells County) had not previously been counted.

The notorious Box 13 gave Johnson 202 votes and Stevenson only one, thus earning a place in history; it sent Johnson to the U.S. Senate by 87 votes. The State Democratic Executive Committee, by a single vote, made the results official on September 14, 1948. The general election was a mere formality in those days, with the Democratic nominees always winning.

Key LCRA employees had been among Johnson's most ardent supporters during that historic primary campaign. And his work with the LCRA, including the soil conservation program, was one of the many factors which contributed to his victory through its appeal to rural voters.

The soil conservation project proved to be successful for several years, according to Gideon, but various state and federal agencies gradually took over that field — and as they did, the LCRA just as steadily relinquished its responsibilities in saving the soil.

With the LCRA's business mushrooming, Gideon finally went to work full-time for the Authority in 1948 as general counsel. Actually, he had been devoting most of his time with the law firm to LCRA matters for about ten years. And no one appreciated more than he did the value not only of the electricity the LCRA furnished but also the flood protection it was providing by then.

> During the 1938 flood, I lived in South Austin and I had to walk across the railroad bridge to get to work. The only other bridges we had were the one on Congress Avenue, which you couldn't reach because of the high water, and the one on Montopolis, which was washed away. A block or so south of the Congress Avenue bridge, where the old Night Hawk Restaurant was, the water was eight or nine feet deep. So we had to rely on the railroad bridge until the water went down. There wasn't much in South Austin then. They were out of water and everything else, and the electricity was off. My wife's folks lived in San Antonio so I took her over there. She stayed about a week or ten days until we got the water and electricity back.

Gideon sees a lot more than flood control, power sales, soil conservation, and recreation when he takes a bird's-eye view of the LCRA's development.

> First, we had to get the Authority started and Wirtz was the moving force behind that. He brought in McDonough, who had been one of the big engineers in New York for the PWA, to head up the construction program and he proved to be an excellent choice. Then they had to sell the power and they brought in Starcke. He did a great job.
>
> And then we had the era where we'd grown so fast and there was such a demand for electric power that we had to get accomplished what the private electric companies said they would never agree to: that is, to allow the Authority to build steam plants. I had that problem, of getting it done and getting them to agree to it, and getting it passed in the legislature — and getting it financed in New York and Washington. Because, if we hadn't done that, the amount of electricity the dams were generating was so small compared with what is now being used . . .

Gaining attention in New York and Washington, as well as

in the rest of the country, received a tremendous boost at the end of 1949, thanks mainly to Wick Fowler and one of his typically zany ideas.

Fowler was an inimitable humorist, newspaper columnist, philosopher, war correspondent, and chili aficionado. He spent much of World War II covering the battles of Texas's famed 36th Division in Europe for the *Dallas Morning News*. (He was credited later with prompting *News* accountants to develop a new expense account form — after he gave them his own version, accounting for his wartime advances by writing on the back of an envelope: "Covering war, $2,000.")

Fowler was serving as editor of a small weekly newspaper called the *Highland Lakes News* in 1949 when he originated the idea of "Operation Waterlift." He described it as a mission of mercy for thirsty Texans stranded in drouth-stricken New York. The publicity stunt focused national attention on the Highland Lakes and their development by the LCRA.

Fowler persuaded the Texas Motor Transportation Association to make arrangements for a tank truck to carry 3,000 gallons of Highland Lakes water to New York. Fowler said that should be enough to quench the thirsts of 100,000 persons and the water would be dispersed primarily to Texans who were "perishing in the big city with hardly a drop to drink."

Signs painted on the sides of the forty-five-foot-long truck proclaimed "Texas Highland Lakes Water for Thirsty New Yorkers" and one on the back said, "Have a Drink on Texas."

Fowler's announcement of the trip on December 21, 1949, said the water would be drawn "from Buchanan, Inks, Travis and Austin Lakes, which have in storage several billion gallons and won't miss this supply."

The *Austin Statesman,* in describing "Operation Waterlift" on December 22, 1949, said:

> Paper cups, printed to show the origin of the water, will be taken along, and after their thirst-quenching service will be kept as souvenirs.
>
> The truck will be stationed at the city hall steps in New York, where all Texans in the area will be invited for a drink. It will then tour the streets.
>
> "The purest water to be found comes from the Highland Lakes," the *Highland Lakes News* said in a story for this week's

issue. "While much of the lakes' displacement is taken up by game fish, many, many billions of gallons of fine drinking water go through the four dams on the Colorado River, generating the hottest and most powerful electricity in the world and irrigating the world's biggest rice crops.

"Why the thousands of Texans now perishing in New York's downtown desert ever left the Highland Lakes in the first place is still a puzzle to this newspaper. But since they are there it is our duty to see that their thirst is quenched.

"A tank truck load may not seem to be much water for a city the size of New York. But since the Highland Lakes water has properties comparable to the fabulous Fountain of Youth, New Yorkers are in for a surprise. This, of course, will not be surprising to the Texans now stranded in the big city."

On December 31, 1949, the *Highland Lakes News* published a special "Water Lift Edition" with a front-page banner headline declaring: "Thirsty New Yorkers Ready for Waterlift." Beneath that was another headline: "Whole Nation Hearing About Highland Lakes." That article read:

Highland Lakes water is splashing over the nation's newspapers, radio networks and newsreels.

In what is probably the biggest spontaneous promotion to come out of Texas in years, Operation Waterlift is nearing thirsty New York from the lakes of the Colorado River. Tuesday is the date set for the New York reception from Mayor William O'Dwyer and his Texas-born bride.

But in the meantime the big 45-foot thermos bottle of pure drinking water is receiving a tremendous reception in the towns and cities through which it passes in Louisiana, Mississippi, Alabama, Georgia, South and North Carolina and on to New York.

Folks all over the United States know about it and those living along the route are turning out for local festivities. At Jackson, Miss., the truck and the Nash advance car were held up by Gov. Fielding Wright. It will stop in Washington for the newsreels.

"I just have to have a drink of that water on my friend, Governor Allan Shivers," Wright told the Texans in charge of the "Mercy Mission."

Gov. "Big Jim" Folsom of Alabama kept the caravan waiting for two hours while he finished a nap . . .

The article noted that in the advance car moving ahead of the tank truck were John Babcock, development director for the LCRA, Charles Ogle, safety director of the TMTA, and Bob Swanson, representing the P.K. Williams Nash Company in Austin. Douglas "Red" Tipton of Houston shared the truck-driving duties with Billy F. Davis of Houston. The article added:

> The group is passing out pamphlets heralding the qualities of the Highland Lakes and having a lot of fun telling the folks about the rumor that a 14-pound black bass got into the tank while it was being filled . . .
>
> Babcock telegraphed back to the *Highland Lakes News* that the truck is being slowed considerably by former Texans begging for a drink out of it. But they told them the water was for Texans stranded in New York City . . .

Throughout the 2,000-mile eight-day trip, the Texas "Mercy Mission" for New York harvested tons of publicity, with coverage by local newspapers, wire services, magazines, newsreels, radio, and the then fledgling medium, television.

A *United Press* article, published in newspapers throughout the country, described the water truck's arrival:

> NEW YORK, Jan. 3 — (UP) — A big water wagon rolled into Manhattan from Texas Tuesday to quench the thirst of "displaced Texans" in water-short New York.
>
> The tank truck, carrying 3,000 gallons of "Texas Highland Lake water," pulled into a midtown parking lot after an eight-day cross country trip from Austin, Texas.
>
> Water Commissioner Stephen J. Carney, accepting the "precious cargo" on behalf of New York, promptly took a drink from one of six handy spigots on the tank.
>
> He smacked his lips and smiled, "Mighty good," he told a crowd of ogling New Yorkers and Texans.
>
> "We appreciate the spirit which prompted the sending of this water," Carney said. "I hope it won't be necessary to truck water into the city. But if we have to, it's nice to know we have friends like you folks."
>
> Mrs. Edward F. Nordell, formerly of Fort Worth, Texas, and now president of the Texas Club of New York, stepped up for a drink.
>
> "It's like a shot in the arm," she said.
>
> The water was presented by Bob Swanson, Austin businessman, who escorted the tank truck in a private automobile.
>
> A crowd of about 100 persons were on hand to welcome the

Texas gift. They included Mary Hatcher and Danny School, of the Broadway musical, "Texas, Li'l Darlin'," and Tex McCrary, radio star and newspaper columnist . . .

All the hype and fanfare apparently left a sour taste, however, in the mouth of the *New York Herald Tribune*, which published a short article in its Wednesday, January 4, 1950, edition about the truck's arrival. The headline declared "Wet Blanket Greets Texas Gift of Water" with this subhead: "City Won't Let Tank Truck Cruise on Streets."

> A group of Texans brought a tank truck full of Texas water here yesterday for a publicity stunt. After the television, newsreel and newspaper photographers had finished filming the event, the Texans found themselves with ten tons of water on their hands.
>
> The original plan had been to publicize the abundant resources of Texas by bringing water to "parched, displaced Texans" in New York who had made the mistake of leaving the Lone Star State. New York police, however, would not allow the truck to cruise around the city with its placards inviting one and all to "Have a drink on Texas," lest it prove a traffic hazard.
>
> So the Texans took the truck to a parking lot at Fiftieth Street, west of Seventh Avenue, in hopes that New Yorkers would drop in for a drink. However, few persons showed any interest, except for the parking lot operator, who demanded a $25 parking fee.
>
> Arrangements were finally made yesterday afternoon for dumping the water into Jerome Park Reservoir, the Bronx, at 11 A.M. today. Jerome Park is a distributing reservoir with a capacity of a billion gallons. The truck brought 3,000 gallons.
>
> The truck arrived at noon on another parking lot, 231 West Fiftieth Street, where the water was sampled by Water Commissioner Stephen J. Carney, members of the cast of "Texas, Li'l Darlin' " and others. The stunt was arranged by the Lower Colorado River Authority, the City of Austin, capital of Texas, and the Texas Motor Transportation Association.

The *Herald Tribune* article failed to dampen the spirits of the Texans, who deemed the publicity stunt an unqualified success. Babcock wrote a report on it dated January 10, 1950, and declared:

> A sizeable block of the people of the United States (estimated conservatively up in the millions) today are conscious of the fact that there is such a place as Austin, Texas, and that at and near Austin are a series of major fresh water lakes nestling

in the picturesque hills, which not only provide opportunities for outstanding recreation, but also contain ample water for development of cities, agriculture and commercial and industrial projects.

Specifically, the people of 12 states, lying in the southern and eastern portions of the United States, have read and heard about the Central Texas Highland Lakes for about two weeks. Uncounted millions elsewhere read and heard about the Lakes at least one time during that period.

That means that for the first time many, many people know there is such an area and that they will be receptive to any further information concerning it. It is easy to believe that everyone has heard of your home town, but you need not get too far away from Central Texas to meet people who don't know of the Lakes, and not too far away from Texas to find people who never heard of Austin and think the Colorado River is in New Mexico.

The major accomplishment of the advertising promotion then is:

The promotion awakened a large segment of the American public to the area.

It resulted in more than 28 newspapers and their staffs, 28 nationwide wire service bureaus, operators of eight smaller radio stations, program directors of all three major radio networks, editors of one major newsreel company and three national advertising agencies being fully conscious of the area. In addition, three major television networks presented information about the area to their audiences . . .

Financiers were among those who discovered the existence of the LCRA through Fowler's highly successful "Mercy Mission" to New York. But ironically, the nationwide attention gained for the LCRA and the Highland Lakes came shortly before the start of a severe drouth in Texas — a drouth which greeted the completion of two new LCRA dams.

The Granite Shoals Dam (later renamed the Alvin J. Wirtz Dam) which created Lake Granite Shoals (renamed in 1965 Lake Lyndon B. Johnson) was completed in 1950. The Wirtz Dam is 118.3 feet high and 5,491.4 feet long, backing up a reservoir 21.15 miles long with a maximum width of 10,800 feet. The lake covers 6,375 acres and has a capacity of 138,460 acre-feet of water when filled to the top, which is 825 feet above mean sea level.

The Marble Falls Dam (renamed in 1962 the Max Starcke Dam) was completed in 1951. It is 98.8 feet high and 859.5 feet long and covers 780 acres, with a capacity of 8,760 acre-feet of water. The top of the reservoir, when full, is 738 feet above mean sea level.

According to the *Houston Post* of June 16, 1952, a crowd estimated at more than 6,000 persons gathered near Marble Falls on Sunday, June 15, for the ceremonies renaming the Granite Shoals Dam in honor of Alvin J. Wirtz.

> . . . Tribute of the highest to the man who made the river authority possible — the late Sen. Alvin J. Wirtz — was threaded throughout the ceremony.
>
> A granite marker, which will be built into the dam, was unveiled by Stephanie Cain, 13-year-old granddaughter of the late senator.
>
> U.S. Sen. Lyndon Johnson, Gov. Allan Shivers, Congressman Homer Thornberry and Tom Miller, former mayor of Austin, were the speakers. Max Starcke, general manager of the LCRA, presided.
>
> The new dam is one of six built by the Authority in a flood control and power development program within the 10-county Colorado River watershed.
>
> Both Gov. Shivers and Sen. Johnson praised Sen. Wirtz as having been the leader in the creation of the LCRA and in building its flood control program.
>
> Gov. Shivers expressed pleasure in the state's being able to sponsor such a project without outside aid, and he pointed out that it was financed from revenue bonds without cost to the taxpayers . . .

The text of the governor's speech included these remarks:

> While he willed us a physical legacy of stone and steel, Alvin Wirtz left something far more valuable — a water conservation program of immeasurable benefit to our own and future generations . . .
>
> We hear a great deal about the importance of oil and gas to Texas — and, of course, their vast importance is not to be discounted. Yet in the final analysis, water is of more value to the economy and the future prosperity of the state. That is why it is so vital that an effective, well-planned and permanent program of water conservation be established, and why it is so encouraging that such a program already has been brought to full flower on the Colorado River.
>
> . . . the LCRA has made possible the irrigation of 53,000 additional acres of Texas rice each year — thereby doubling our rice

production. It has increased Texas farm income by $50 million in the past ten years. It has saved the state an average of $2 million a year, or $20 million in the past decade, in flood losses.

These are impressive accomplishments. And the greatest source of pride and satisfaction is that LCRA is Old Man Texas' own baby — born in Texas, nurtured in Texas, owned and operated by Texas with no strings attached. LCRA is a perfect answer to the charge so often made that all government is wasteful and expensive. It is proof that a state agency can be operated in an efficient, economical and even profitable manner.

For example, this Dam that we are today dedicating to the memory of Alvin Wirtz was built with revenue bonds — financed without a penny's cost to the taxpayers. The total cost of the Dam and power house, I am told, was nine million, 825 thousand dollars. Its counterpart at Marble Falls cost more than seven million. Together, they add about seventeen million dollars to the physical properties of the state, not to mention the inestimable value derived from water conservation, flood control, power production and recreation. All of this has been achieved without the addition of one thin dime to the tax bill.

In water conservation alone, the system of six dams and lakes has made a most outstanding contribution to the economy and the welfare of Texans of today, and those of tomorrow . . .

Senator Johnson's remarks focused more on his personal relationship with Wirtz, his longtime friend, upon whom he lavished extravagant praise.

He was the humble son of a manual laborer. He himself worked first as a day laborer — faced with no future except what he himself might make.

In those boyhood years, on the banks of the Colorado, Alvin Wirtz came to know this river well. He knew that it was the hope of all who lived along its flood plain — and, at the same time, the most feared enemy of their tranquility and their accomplishments.

He saw the river, in its placid moods, bring prosperity to the farms and the cities and villages along its banks. He saw the river, in its angry floods, bring destruction and death and privation to those who trusted it.

In those years, there was born in Alvin Wirtz a determination to master this river — to make it a servant of the people who were forced to trust it.

In his lifetime, he served in many posts. He rose to the highest councils of his nation's government. He was known, respected, and trusted — and his advice was sought — by Presidents, Supreme Court jurists, Cabinet officials, members of Congress, leaders of business and industry and labor.

Through all that time, his head was never turned. He never lost sight of his original purpose. Wherever duty might call him temporarily, he always came back to the valley of the Colorado to carry on his fight for mastery of the river . . .

Johnson went on to say that Wirtz, while serving in the state Senate, saw the beginning of his dream — a string of dams along the Colorado. The Depression shattered that dream, he noted, but not Wirtz's faith in his fellow man.

Standing almost alone, while others wailed and wept that all was lost, Alvin Wirtz began to talk and argue and plead and work. He succeeded in convincing the people of the Colorado Valley and, eventually, his former colleagues in the legislature, that the Colorado could be harnessed — and that they could do it.

He secured passage of the legislation enabling the Lower Colorado River Authority to come into being. From that point on, he was relentless in his endeavors . . .

Alvin Wirtz was — as so many of you know — my dearest friend, my most trusted counsellor. From him — more, perhaps, than from any other one man — I gained a glimpse of what greatness there is in the human race; how, when they are free, men and women can and will find their own salvation and rise to meet any challenge laid before them . . .

This structure of steel and concrete cannot recapture the warmth of his friendliness, the strength of his character, the inspiration of his vision. But it can stand as a reminder that one man — who believes devoutly in the people — can work greater wonders than hundreds of thousands of men who have no such faith . . .

Throughout the eloquent speeches of Shivers, Johnson, Tom Miller, and Congressman Thornberry, who succeeded Johnson in representing the Tenth District, one young man cringed constantly in the fear that their words might suddenly begin falling on what would amount to deaf ears.

J. J. (Jake) Pickle, who would succeed Thornberry as the Tenth District Congressman in 1963 (when Thornberry was named to the U.S. Fifth Circuit Court of Appeals), served as coordinator for the dedication arrangements. That included working with Marble Falls volunteers who served barbecue and handled parking that day, making detailed plans for seating 3,150 persons in a rented, circus-type tent 200 feet long and 60 feet wide — and arranging for the public address system.

Almost everything went smoothly. With eight efficient serving lines, the Marble Falls volunteers even managed to serve 7,672 plates of barbecue in just twenty-seven minutes. In retrospect, Pickle wished they had been a bit slower, because suddenly it was time for the celebrity-studded ceremonies to begin. Pickle recalled twenty-three years later:

> Senator Johnson had told me to be sure and set up two public address systems because he said "something always goes wrong with one." I got Moton Crockett to set up a system and decided that another one, which would be rather expensive, was unnecessary.
>
> Sure enough, that thing went on the blink just before the dedication ceremonies were supposed to start . . .

Crockett also has vivid memories of that problem. For some reason he didn't carry extra equipment with him that day, though he usually did. "I had to race all the way back to Austin to get some more. So we had a delay of about thirty minutes in getting the program started."

Pickle said Johnson spent most of that time chewing him out, reminding him over and over again that he had told him specifically to have two systems set up.

"It was," Congressman Pickle said ruefully, "the longest thirty minutes of my life."

19

Getting Steamed Up

Gideon described the history of the LCRA as "one crisis after another during the earlier years." After World War II there was a period of relative calm, he said, but that was disturbed by the fuel crisis which fell upon the country.

Between World War II and the fuel crisis, the LCRA faced the problem of winning legislative permission to build the steam plants necessary to generate the power its customers needed.

According to Gideon, the process went slowly.

> We went in and visited with the utility company people and convinced them, finally, that we had to do it — but we were not going to go into competition with them. So each time we needed to build a steam plant — and the first one was down at Bastrop — we'd have to get a law passed. When we wanted to increase the capacity, we'd have to get a new law passed because the first one limited us to a certain capacity.
>
> In order to serve what we were serving, we had to have the ability to build steam plants. We would have been willing to buy power from the private utilities but they wouldn't sell it to us. They said they were not going to build steam plants to sell us power because we could build them as cheap as they could. So it

184

was just a question of working it out. But it was quite a change. We were prohibited from building steam plants and they said that was one thing we'd never be able to do. But we worked it out every time and when we did, we always got a pretty good vote in the legislature.

One thing we had going for us was the fact that the electric rates of the Authority were so low that there wasn't any reason for anyone to stir things up. Nobody wanted to go back to the higher rates. Also, the LCRA dams had stopped some floods — and had provided water during drouth conditions. So we were getting along pretty well.

Starcke retired on December 31, 1955, and Gideon became general manager on January 1, 1956.

Gideon generally is credited with "running a tight ship" during his tenure as general manager. To some, it was even too tight. He had two assistant general managers: Lucksinger, the controller, and G. E. Schmitt, the chief engineer. Some employees resented the fact that Gideon believed strongly in a chain of command and prohibited them from talking with LCRA board members — with whom some of them had been on a first-name basis for years. It was a far cry from the old "family type" atmosphere that had prevailed under Starcke. But Gideon felt the board should set policy and not even be tempted to get involved in the day-to-day operation.

In addition, a tight money situation as postwar inflation heated up kept many LCRA employees from getting badly needed pay raises while their own household expenses were rising rapidly.

Soderberg recalls going about two years without a raise while he was working in the meter shop. But he understood the Authority's financial problems.

"A lot of times," he said, "we couldn't even afford to buy parts. We'd have to make them ourselves — things like coils, wires and so forth . . . Part of it was because our rates were too low. We couldn't adjust them fast enough to keep up with inflation."

The LCRA's problems were not entirely of a financial nature. One of the worst drouths in Texas history plagued the state from 1952 until 1957. In the Colorado River country, it really began several years earlier and then paused for massive rainstorms on September 10 and 11, 1952. The rains came while Lake Travis was at the lowest level in its history; the lake rose more than fifty feet in less than twenty-four hours and went up a total of fifty-seven feet before the storm stopped. Mansfield Dam thus prevented disastrous flooding in Austin, where engineers estimated the Colorado would have reached the forty-five-foot stage. That would have been four feet higher than the record-breaking flood of June 15, 1935, which caused property damage of approximately $13 million.

In a speech at Bay City celebrating "Rice Day" on February 4, 1953, Starcke told how he learned of the heavy two-day rains (15.76 inches officially at Fredericksburg, for example, and 14.73 inches at Llano). He said he and his family were in Lucerne, Switzerland, on his first vacation in fourteen years.

> At 12 noon — Swiss time — the phone rang in my hotel room. In broken English, the operator told me that Mr. Sam K. Seymour of Columbus, Texas, was calling.
>
> "Max, Max" he said excitedly. "We've got a big flood on the Colorado. All the lakes are full. It's still raining. Why, one place has had twenty-seven inches. A lot of roads are washed out. Six bridges washed away . . ."
>
> I think I said, "Why, that's fine!"
>
> When you have had our experience of spending seven or eight years seeing the reservoirs on the Colorado going down six, seven, ten, fifteen feet a year — while the Texas sun keeps sucking up water . . . And when you have spent sleepless nights worrying about how you are going to have enough water to take care of the needs for rice irrigation, and at the same time take care of the municipal and industrial users of water — and to make the electricity that pays the bills of the LCRA; well, just put yourself in our place.
>
> July of 1945 had been the last time that Lake Travis, our biggest reservoir, had been up to the so-called operating level (681 feet). Water requirements by all users had caused this lake to go down, in 1952, to its lowest level since its original filling.
>
> We all had come to the conclusion that it just wasn't going to rain any more in our watershed. We had even gone to the extreme of inviting the Krick organization to try their hand at a rain-increasing program. But after six months, they, too, had become discouraged and packed up and left our area.
>
> I wasn't doing the rain situation any good in Austin, so I took the trip. And there I was, halfway around the world . . .
>
> Since I've been home, the board of directors of the LCRA has facetiously suggested that if my going to Europe would

make it rain in our watershed, then they had better send me every year . . .

Starcke did not go back to Switzerland but he had retired before that drouth ended. It finally was broken with a vengeance. During a sixty-three-day period from mid-April until mid-June of 1957, the Colorado watershed was battered by eight king-sized thunderstorms which dumped a total of three million acre-feet of water into the river. The LCRA dams were credited with preventing floods which would have caused an estimated $35 million in damages. The actual damages inflicted by those torrential rains totaled approximately $200,000.

Throughout the 1950s, the LCRA was able to meet its demands for electricity with hydroelectric power supplemented by the small, gas-fired Comal Power Plant it leased in New Braunfels. But as the Central Texas population grew and industry began rushing into the area, it became obvious that those sources would not be enough.

Gideon quarterbacked the "gentle persuasion" which pacified the privately owned utility companies enough to permit passage of a bill authorizing the LCRA to build its first thermal generating plant. Governor Price Daniel signed the measure on January 31, 1962. It included many restrictions but basically permitted the Authority to build a plant of no more than 250,000 kilowatt capacity; its hydroelectric power capacity then was 230,000 kilowatts.

A year later, on January 23, 1963, the Rural Electrification Administration approved an $18 million loan for construction of the Sim Gideon Steam Plant near Bastrop and for a network of transmission lines to distribute the power it would generate.

Approval of the loan was announced jointly by U.S. Senator Yarborough and Congressman Thornberry, who credited Vice-President Johnson with an assist in obtaining it. The agreement climaxed year-long negotiations involving the REA, the LCRA, and the Central Texas Electric Cooperative, Inc., headed by President Troy Foster and Manager W. C. McWilliams.

Gideon noted, in commenting on the loan, that Johnson had said repeatedly during negotiations "he wanted the people of Central Texas to continue to have the social and economic benefits which had come to them for more than two decades as a result of low-cost electricity furnished by the LCRA."

A direct result of the REA loan for the new generating plant,

Gideon said, was that the people of Central Texas would save $1,950,000 in the cost of electricity over the next twenty years.

The announcement of the loan came eight days after former secretary of the navy John B. Connally, who had served on the LCRA board from 1941 until 1947, became governor of Texas. Connally, one of Johnson's closest friends and allies for many years, was inaugurated on January 15, 1963. Most of his time as a member of the board had been spent on active duty in the navy during World War II, but his actions as governor were destined to have a profound impact on the LCRA.

Those actions also illustrate the diverse factors which affect gubernatorial campaigns and the appointments made by Texas governors to various boards and commissions. Connally's relationships with two men who would play key roles in LCRA history, plus their association with each other, provide convincing evidence that politics not only makes strange bedfellows; politics, sometimes, produces strains on friendship that result in surprising and unusual sets of enemies.

Bill Petri, a sometimes brash but always unabashed labor leader, was working as a stereotyper in the mechanical department of the *Austin American-Statesman* when Charles F. Herring first ran for the state Senate in 1956. Petri and Charles E. Green, the editor of the paper, were close friends despite their conflicting viewpoints on labor matters. Green asked Petri to support his friend, Herring.

Petri said he did not know anything about Herring, a World War II navy hero who had served as an aide to Congressman Johnson and later as U.S. attorney for the Western District of Texas. But after analyzing the race, he decided Herring would win it and stuck his neck out for him. That infuriated many of his colleagues in organized labor; however, when Herring won, Petri represented the unions' only contact with the new senator.

Petri and Herring became close friends. Petri later told this story:

> I was in Herring's office one day in 1962, and he told me Connally had decided to run for governor. The phone rang and it was John. He was still Secretary of the Navy and was calling from Washington. Herring said, "Yes, Petri's here right now." He told me Connally wanted to talk to me. Connally said he was coming down the next Monday to announce for governor and said, "Not just for me, but because of your closeness to Herring, I hope you can support me."

I told him I had only one question to ask: "Are you coming down here to serve one term and then run against Ralph Yarborough for the Senate? If you are, the answer is no!"

Connally said no, he was not going to serve one, two, or even three terms and then run against Senator Yarborough. I told him to call Yarborough and tell him that, and to tell him he had just talked to me. I was trying to help Yarborough but what do you suppose Yarborough told him when he called? I didn't know this until about two months later but Connally called him, told him he had talked to me and I had asked him to call and assure the Senator he did not intend to run against him later. Yarborough told him he didn't care and for him to come on down, get elected governor and "you still couldn't beat me." The next time I saw Yarborough, I collared him and asked why he had made that statement. I told him it was not smart — that Connally didn't call him because he wanted to but because I insisted.

Connally won that race and then — for some reason — turned against Charley during his first legislative session, in 1963. They got cross-ways, I think, because Connally was jealous of Herring's closeness to the blacks, Chicanos, and labor.

But regardless of their personal relationship, Herring had the right as a senator to veto anyone from his district whom the governor might want to appoint to a board or commission when Senate confirmation was required. This is an ancient "senatorial courtesy" tradition, a customary trade-out between senators which has survived many perilous pitfalls. And Connally had to appoint two persons from Travis County, in Herring's district, to serve on the LCRA board.

Connally was in his second term as governor when he called Petri in and officially offered him one of those appointments while making it clear he had not originated the idea.

As Petri tells it, Connally said:

"Look, you and I don't get along but I've been blackjacked. I've been told by your buddy, Herring, that I had to appoint you or he'd kill the appointment of anyone else I nominated."

Connally told me that it was a pretty good board and that he had served on it. But he said there were a bunch of old people on it and most of them were rich. He said since I had served on the Austin Housing Authority Board he knew I had a knack for getting along with people like that. And he said I would have to, because there were a lot of things that needed to be done. As an example, he said there was lots of open land belonging to the LCRA and the public ought to be able to enjoy it. He told me the LCRA had $38 million in Austin banks that ought to be used to develop that land into parks, so everybody in the state could enjoy it.

Connally did not recall the details of that conversation during a 1986 interview but declared that he always had tried to appoint outstanding individuals to all boards, and that he had special respect for the LCRA. Said Connally:

I think the LCRA is a classic case of what could be done through a public institution that private industry would not have done for generations, if ever. Without the LCRA, in my opinion, you could never have had the coops — because they basically got their power from the LCRA. At least, they would have been much slower getting started down here because this was a real battleground between public and private power. The utilities would have been extremely slow to see the coops develop and to sell them power.

Lyndon Johnson was in the forefront of the battle, of course, and he instigated the formation of the coops. We went to a lot of meetings to organize those coops — and some of those meetings were held in schools where we had to use coal oil lamps. He used that to help convince the people they needed electricity.

And this proved to be politically popular because the beneficiaries of rural electrification were so many, and they were so grateful. It had such a profound effect on their lives that they never forgot it. On the other hand, people in big cities like Austin already had electricity and they were not worked up. It was a non-issue so far as they were concerned . . .

It was indeed a politically popular thing — because it was one-sided, so far as the numbers were concerned. It was something really worthwhile that was also politically acceptable. It just changed the whole quality of life for people.

And I sensed it so much, because I grew up down in Wilson County. We moved back there in 1932, to a farmhouse that had no electricity, no running water and no indoor plumbing. I graduated from high school studying by kerosene lamps. All my brothers and sisters did, too. My mother cooked on a wood-burning stove until after I was out of the University of Texas. We didn't get electricity down there until 1940.

I had graduated from the university and gone to Washington. The other children and I bought my mother an electric stove for Christmas in 1940. It was the first electric stove she had ever had. Wood-burning stoves, coal oil lamps and lanterns — that's what we all lived by and studied by and worked by. It's hard now to believe that in our lifetime we went through that . . .

Petri said Connally asked him not to say anything about his appointment until it was announced along with three others then pending for the LCRA board. Several weeks later, Connally announced his appointments of Tom Miller, Jr., son of the former Austin mayor; Judge Ferguson, who had served on the board from 1935 until 1937 and again for six months in 1945; E. A. Arnim of Fayette County, who was being reappointed for his fourth consecutive six-year term; and Petri. The new appointments became effective February 5, 1966.

Right from the start, Petri and Gideon seemed to clash. Petri recalled during a June 1985 interview:

> Before my first Board meeting, Gideon told me how rich most of the other members were and said he wanted to make a suggestion to me. He said some of those board members had been sitting in the same chairs for twenty-five or thirty years and if I sat in one of their chairs they might be upset. I said: "Oh, hell, don't worry. I'll wait until they all sit down and then take the last chair. And if there's not one left, I'll sit on the floor. But Sim, before I leave this board you're going to know that I was on it!" He was just trying to embarrass me because I came from organized labor. But I did just like he said, and there was a seat left so I took it.

Mac Umstattd, who served as general counsel for the LCRA from 1958 until he retired in 1983, witnessed many of the sparks that flew between Gideon and Petri. And he respected both men, he said.

> I don't think Gideon tried to embarrass him consciously. Gideon was basically shy. He was not as much at ease in handling people as Mister Starcke had been. He was a brilliant man — a very caring person, kind and compassionate. But his emphasis was always on the job and getting the job done.
>
> Because he had been a poor boy, I think subconsciously he favored the wealthy members of the board — and several of them were millionaires. Petri, as a labor negotiator, had something of a chip on his shoulder. They were both products of their environments but both were operating in good faith.
>
> And Petri did more than anyone else to get LCRA salaries up. Lucksinger was controller and he had worked for twenty-five

cents an hour at Buchanan; a dollar was a dollar to him. Gideon was pretty much the same way. Lucksinger ran a very conservative structure and he was pretty put out with Petri for trying to get the salaries raised. Lucksinger was inclined to hold the line on any expenditures.

Petri waited until he had been to two or three board meetings before he began speaking up; then he did so in no uncertain terms, which sometimes ruffled the feathers of his colleagues. But he was pleased that the members finally began debating the issues.

Petri recalls with pride a remark made to him by Sam K. Seymour, Jr., of Columbus, who still holds the longevity record — thirty-six years, from 1945 until 1981 — for service on the board.

> He told me once that before I got on the board it was always a unanimous vote and "now we're all torn up." He was trying to make me feel bad. But I told him: "Mister Seymour, you've got to be kidding. You were proud that it was always a unanimous vote?" He said, "Yeah." And I said, "Well, for twelve men to always be in agreement, something was wrong. It had to be. I'd be ashamed to tell it."
>
> Herring and I usually had lunch about once a week. I told him it was disheartening and asked him if there were boards like this all across the country. He said, "Hell, you're on one of the good ones." I said, "No wonder this country is screwed up. If this is a good example of people who are leaders in their community, serving on a board that's just a rubber stamp for the general manager — this scares hell out of me."
>
> But Connally had told me the board members were good people and that I had to get along with them or I'd be of no effect. Of course, I already knew that but I appreciated his telling me that.

Petri said he became especially upset after he had been on the board for a few months and found that the employees, many of whom he had known for years, would not speak to him. He finally asked one of them what the matter was and his friend said, "Gideon, Schmitt, and Lucksinger; they'll nail me if they catch me talking to you."

Petri went directly to the three men and made it clear that he was going to start talking to everybody there and that they had better not cause any of those employees any trouble. "I also had some kinfolks working there that they didn't even know about," he said. "I was determined I was going to find out what made the place tick. And since I was working nights at the *American-Statesman,* I had plenty of time to spend at the LCRA during the daytime."

In many ways, Petri was pleasantly surprised at what he learned.

> It's a wonderful organization. It has done a lot of good . . . in spite of itself. The most important thing you've got is the water — it's far more important than the electricity. You can make more electricity but they ain't making any more water. And the LCRA also has kept the private utilities honest.
>
> Gideon, Schmitt, and Lucksinger had a kingdom out here and they ran it just like they wanted to. I'm not saying they didn't do a reasonably fair job — except that I don't think they treated the employees properly. Morale was very low — and so were salaries — when I came on the board.
>
> If anything has been rewarding to me, it was that I was given an opportunity to do some things that I wasn't able to do out in the private sector, as a labor official. I was able to do it here real easy. It took five or six years but I'm seventy-two years old and in my life, that's not a very big span.
>
> Where I'd been negotiating raises for fifty years and getting little, pitiful raises, I was able to come in here and the same people that were making $7,200 a year — the women — now some of them are making $30,000. All of the discrimination against the women was eliminated. My mother was the best thing that ever happened to Bill Petri. And she was strong. I've got a good wife, a daughter and granddaughters; there's no way I could sit up here and see women discriminated against . . .
>
> And Herring gave me the opportunity. I got on the board strictly because I had helped him in that '56 campaign.

One of Petri's big disappointments was his failure to win board approval for Connally's recommendation that some of the LCRA's lakeshore property be developed into public parks.

"I tried," he said, "but the other board members insisted they were not in the park business. They have many, many miles of lakefront that could be developed. I tried to get 'em to develop the Pace Bend Park but they finally turned that over to Travis County."

Petri was destined to serve fourteen years on the board, including two as chairman, and Herring eventually would spend eight years as general manager of the LCRA. Ironically, they were serving in those positions when their close, long-standing friendship disintegrated.

20

Gas
Attacks

In 1951 the LCRA completed the smallest of its six dams, near Marble Falls. The dam creates the smallest lake in the Highland Lakes chain, only 5.75 miles long with a maximum width of 1,080 feet. For eleven years, both the lake and the dam bore the "Marble Falls" name, but that changed on October 31, 1962. Wray Weddell, Jr., described the dam's name-changing ceremony in the *Austin American* on November 1, 1962:

> MARBLE FALLS — The Colorado River dam here is 90 feet tall and contains more than three million pounds of steel and iron but it is no bigger in the eyes of Texas water conservationists than the man for whom it was named Wednesday.
>
> When the day dawned bright and chilly, the massive concrete structure blocking the river's flow through a rock-ribbed gorge was known simply as the Marble Falls Dam.
>
> Shortly before noon, it was the Max Starcke Dam — so dedicated in an hour's ceremony as lasting tribute to the man, short of build but tall of achievement, who headed the river-harnessing Lower Colorado River Authority for 17 *[sic]* eventful years.
>
> Men in high places devoted to saving precious water and making it work for the better life were here to say as best they knew how that the choice of name was as it should be.

194

Declaring this were Governor Price Daniel, Congressman Homer Thornberry, LCRA Board Chairman Sam K. Seymour of Columbus, Texas Water Conservation Association General Manager J. E. Sturrock, and Sim Gideon, who succeeded Starcke as LCRA general manager.

And from Vice President Lyndon B. Johnson came a letter of tribute and congratulation.

Starcke, who will be 79 on Nov. 11, responded with a single softly-spoken sentence, ending with "God bless you."

Minutes later Starcke told *The Austin American:* "Usually these things come to people after they have gone to their eternal reward. I'm happy to be here."

The ceremony scene was on the parking lot of the Starcke Dam power plant perched high above the foam-covered spillway — picturesque, but no place of comfort for those afflicted with fear of height.

The honor of unveiling the dedicatory plaque on the power plant wall went to Starcke's attractive 20-year-old daughter, Mrs. Margaret Woodruff of Austin . . .

Daniel, an insistent advocate of flood control and water storage, said of the agency which Starcke led as general manager from 1940 until his retirement Jan. 1, 1956: "The work on the LCRA has been a model — not only in our state and nation, but in other nations."

Turning to Starcke sitting nearby, Daniel said: "The State of Texas owes him a great debt, as do Austin, Seguin, and many other cities and the countless organizations to which he has given his time and energies . . ."

Similar tributes came from other speakers. Said Thornberry: "It is particularly fitting that this magnificent dam with its sinews of steel, its ponderous weight of concrete, its powerful electric generators, should be named to commemorate the life and works of Max Starcke — the vision, the humanity, the tenacity and remarkable skill of this outstanding Texan."

Seymour termed the dedication "a fitting and eternal tribute to a great man — a memorial to what one man's abiding faith in the human race can do."

In sharp contrast with the Starcke Dam dedication, the Gideon Steam Plant near Bastrop began operating on May 15, 1965, quietly and without fanfare. That, recalls Soderberg, was the way Gideon wanted it. Open houses were held at the plant October 28–31, 1965, but received little publicity. The $18 million project included a dam 4,500 feet long, creating a 960-acre reservoir, a pumping plant, and a 3.5-mile-long pipeline along with a 125,000-kilowatt, gas-fired generating plant. Another 125,000-kilowatt unit was scheduled to go on stream in 1968.

Even less attention had been focused in 1962 on another event that turned out to be momentous: the LCRA's signing of a twenty-year contract with the Coastal States Gas Producing Corporation, headed by Oscar Wyatt, for fuel. Deliveries were to begin in 1964.

In a 1980 interview with John Williams, editor of the LCRA's *River Review* magazine, Soderberg recalled:

> At that time, we felt we had a good gas contract, very inexpensive gas costs — I guess you were talking about 20 cents per thousand cubic feet (MCF) — and supposedly an ample, good supply. Even with the addition of construction and fuel costs, LCRA was producing some of the lowest-cost electricity in the United States. We were considered the good guys with white hats, and everyone seemed to be happy and pleased with our progress.

The honeymoon with Coastal States lasted long enough for the LCRA to build two more gas-fired units at the Gideon Plant and start construction near Marble Falls of what would become known as the Thomas C. Ferguson Power Plant.

Meanwhile, Petri had been working to get his friend, Senator Herring, named general manager of the LCRA. Despite what many board members considered his abrasive manner, Petri was elected secretary of the board in 1968, 1969, 1970, 1971, and 1972, then chairman in 1973 and 1974. He was the first person ever elected chairman for a second term; after that, he was elected secretary again for 1975. His long tenure as secretary, he says, was due largely to the fact that the board needed someone who lived in Austin to fill that post so he would be handy to sign papers.

Petri said it was shortly after Herring won a tough reelection campaign to the Senate in 1968, over Representative Wilson Foreman, that he first expressed a desire to become general manager of the LCRA. "He told me . . . that he didn't want to go through another campaign. 'I've bled enough,' he said."

Petri said he immediately began exploring the sentiments of LCRA board members. He added that he and Herring also began "lobbying" with Governor Preston Smith to get new appointees named who would be inclined to favor the idea of employing Herring.

That is not the way Herring tells the story. He said:

The LCRA people came to me several times, said they wanted to get rid of their general manager and wanted me to take the job.

I got reelected to the Senate in 1972, but after the regular session of the legislature in 1973, I was really tired and disgusted. I was fed up and ready to leave the legislature.

I had talked with Connally about appointing Petri to the LCRA Board — that's true. I said he represented organized labor but had worked in all my campaigns. But I didn't "blackjack" John Connally. Nobody could do that.

Although Herring never admitted campaigning for the general manager's job, some of his friends say his efforts to gain it were obvious.

John Hancock of El Campo, a banker and rice mill owner who was appointed to the LCRA board in 1963, said during a 1987 interview that he had known Herring all his adult life.

I knew him before he married Doris Wallace. She was from my home town. I talked to Charley and asked him if it was true he wanted to be general manager. I said, "If you do, tell me. I'm your friend; I might be able to help you." He never would admit it. But all the time, he had Governor Smith appointing his people. I know of instances where they actually extracted a promise that if they received an appointment, they would vote for Charley.

I accused one of them of that, and he denied it. But, bless his heart, by the time I got back to El Campo, he was on the phone, crying. He said, "I never lied to a man before in my life — and I lied to you."

Soderberg says it was common knowledge that Petri was trying to get the general manager's job for Herring:

The first time I ever met Herring was at a reception the board had in a private club in Austin four or five months before he was named general manager. Petri was chairman of the board then and I think this was held primarily so the board members could meet Herring. All the top officers of the LCRA, including Gideon, were there.

Petri had talked to me privately on a number of occasions and said he was going to get Herring out there. He said he was a super man for the general manager's job with his experience and his knowledge of the LCRA, the water situation, legislative actions and everything. He felt he was the ideal man to take over.

Herring's law firm — Small, Herring, Craig, and Werkenthin — had represented the LCRA on many matters for years. Clint Small, Richard Craig, and Fred Werkenthin were the other partners. But Small had represented Coastal States when it signed its twenty-year gas supply contract with the LCRA in 1962, and this would cause a slight complication later when the LCRA took Coastal States to court.

Governor Smith gained a majority on the twelve-member board early in 1973 (the membership was increased to fifteen members in 1975). Smith's appointees included Aubrey Voekel, a Fayette County banker, and Roger Gilbert Zercher, a Blanco County farmer, both of whom took office on March 2, 1971. He named Robert Long of Austin, a former Travis County district attorney, to a term beginning November 4, 1971. Smith appointees who took office on January 4, 1973, included Tom Dean, a rancher from San Saba County; Charles Jungmichel, an insurance man and former state representative from Fayette County; Eli Mayfield, a lawyer from Matagorda County; Cecil Long, a Bastrop County merchant, farmer, and rancher who was reappointed; and Petri, also a reappointment.

Minutes of the LCRA board meeting held on May 24, 1973, show that Herring was chosen unanimously to become general manager, effective June 1, 1973, at a salary of $55,000.

Others serving on the board at that time were M. C. Dalchau, a Llano merchant; W. D. Corder, a Burnet businessman; Seymour and Hancock. According to the minutes, Corder was not present at the May 24 meeting.

Herring took office on June 1, 1973 — his fifty-ninth birthday. He was another LCRA leader who knew what it was to live without electricity. Raised on a farm near McGregor, he said there was no electricity in his house until he was a senior in high school.

> We had coal oil lamps. The first time we turned on a light, I thought heaven had opened up. It was hanging down on a cord. We got it through a coop and it was more light than I had ever seen.
>
> Because of federal subsidies, our light bill ran only three or four dollars a month. But you didn't have anything but lights. You didn't have dishwashers and refrigerators and electric irons. You don't use much electricity with just a light bulb.

Herring graduated from McGregor High School in 1931, when

he was sixteen years old, and his father gave him thirty acres of land to farm, along with the mules to do the work. Herring picked his own cotton and saved every dollar he could. After two years he had $200 — enough to get started at The University of Texas.

"I took in laundry and sold sandwiches and went to school winter and summer. It was a lot easier than picking cotton — and I got my law degree in 1938," he said.

Herring recalls that he first met Lyndon Johnson in 1937.

> I was in law school when he ran for Congress in that special election. I was interested in politics, as most law students are. So I decided I'd go out and help him. I put up posters and that type of thing, as a volunteer — and I did get to meet him. After I got my law degree, I went to work for the Looney and Clark law firm in Austin — and Johnson was a good friend of both Everett Looney and Ed Clark.
>
> I was involved to some extent in Johnson's 1941 campaign, when he ran against Gov. W. Lee O'Daniel for the U.S. Senate and got beat. Then I went to Washington and worked in his congressional office for about a year. I volunteered for the navy and Johnson didn't know it until my orders came in. Then he tried to talk me out of going in — he said my job was there. But I spent two years in the Pacific during World War Two and I'm not sorry for it. I made nine assault landings, and had two boats shot out from under me before I got out in December of '45.

Herring went back to work for Looney and Clark, he recalls, at a salary of $250 a month.

> When Johnson decided to run for the U.S. Senate again in 1948, he asked John Connally and me to run his campaign. The firm let me off from April 'til November and we operated out of the old Hancock House in the race against Governor Coke Stevenson.
>
> And we won overwhelmingly, of course, by eighty-seven votes. I spent a lot of time with Mister George Parr of Duval County — especially right after the election, when some of the boxes were so slow reporting results. He used to call me every morning and ask, "How many more votes do you need?"

Herring also was involved in several lawsuits that resulted from that controversial election.

In 1951 he was appointed U.S. attorney for the Western District of Texas and held that post until 1955, when he resigned and began planning a 1956 campaign for the Texas Senate.

I went door-to-door during that 1956 campaign, but of course there was no way I could cover the whole town doing that. So I got six or seven law students, all good-looking boys, who would go up and knock on doors and say, "I'm Charles Herring." Of course, I was a lot younger then so that made it easier to do — and we didn't have TV to worry about much in those days. That way, we made just about every house in Austin.

During his sixteen years in the Senate, Herring sponsored many bills for the LCRA in addition to representing it in many legal matters. He learned a great deal about its operations but none of that knowledge prepared him for the crisis he faced the day he became its general manager.

Storm clouds had been gathering for months. Natural gas supplies began dwindling and, as a result, natural gas prices rose rapidly. Coastal States had advised its customers that it was assigning their contracts to a subsidiary, LoVaca Gathering Company, but the LCRA refused to accept the transfer. And although LoVaca technically was supplying gas to the LCRA, it was Coastal States that the LCRA went after when the firm began curtailing gas supplies from time to time.

Coastal States/LoVaca went to the Texas Railroad Commission early in 1973 seeking permission to breach its contract with the LCRA and other customers by charging higher prices, contending this was necessary in order to obtain adequate gas supplies.

"Coastal States," said Soderberg, "was saying it could not compete in the open market at the prices stipulated in our contract — that they could not go out and buy gas and they were short of gas. They said they could not be competitive unless the Railroad Commission would grant them some relief and allow them to raise the price enough to take care of the difference between what they were having to pay and the price in the contract."

After a series of hearings, the commission granted permission for Coastal States/LoVaca to "pass through" to their customers the higher charges for gas as a "fuel adjustment charge."

Soderberg recalls vividly one of the early side-effects of that problem which added fuel to the fires of the long-running "Texland" controversy. Gideon, he said, called a meeting of all the LCRA's wholesale customers, including the Bluebonnet and Pedernales coops. At that time, those two were buying about twenty-four percent of the LCRA's electricity.

A. W. Moursund, a Johnson City lawyer and close, longtime friend and business associate of Lyndon Johnson's, was serving as general counsel for the Pedernales Co-op. According to Soderberg:

> Gideon told them that we had to do something about our contracts with them because they had a nine-mil limit on them — and Coastal States was having to pay more than that for fuel. Moursund got up and said, "Hell, no, we're not going to change those contracts — you ought to hold Coastal States to their contract!" There was a bitter exchange at that meeting. Moursund and Clint Small, who was representing us, almost came to blows. Moursund left in a big huff and he's been mad ever since.

Small grins now as he recalls the incident.

> We had some rather harsh words. Moursund had become a bitter enemy of the LCRA. Frankly, he thought he was bigger than the Authority. Nobody could get along with him. Herring tried, after he took over — and if anybody could have gotten along with him, it would have been Herring — but he couldn't. Moursund was trying to break up the system.

Small said he was shocked by the intensity of Moursund's opposition as a lawyer. Soderberg, however, was not. He said:

> The Pedernales and Bluebonnet coops always had it in the back of their minds to pull out of the LCRA. Back in the '60s, before we built the Gideon Steam Plant, they wanted to build their own plant. Lyndon Johnson was involved in that fracas. He told the two coops to stay with the LCRA and he would do everything in his power to see that the LCRA got an REA loan at two percent interest to build a power plant. Moursund was working for Johnson at the time and he was instrumental in helping us get that loan. But I heard that the final blow that really ticked him off was when they named that plant the Sim Gideon Plant, instead of the A. W. Moursund Plant. That always stuck in his craw because he felt he was more involved in getting the funds for it. At least, that's what I heard.

Soderberg's promotion to chief engineer became effective the same day Herring took over as general manager: June 1, 1973. Herring remembers the day:

> Elof came in that morning and said, "We've got problems. We don't have any natural gas." I asked him how much we needed and he said, "a bunch — millions of cubic feet. We don't have

any." We had fuel oil to use in emergencies but he said we had only a three-day supply of that. And we'd buy some more but there was none available — so the whole system would go out in three days.

Coastal States and LoVaca had sold more gas than they could supply; darkness, it seemed, was about to settle over much of LCRA land. But Herring's political contacts paid off. He called Bob Herring, chief executive officer of Houston Natural Gas (no relation), who agreed to sell the LCRA fifty million cubic feet of gas a day, beginning the next morning. But Coastal States also had curtailed its deliveries to the cities of Austin and San Antonio. Charles Herring recalled:

> I told Bob Herring that both cities had the same problem we had because they were on Coastal States, too. We had a meeting in their boardroom in Houston. He came in and said: "Let me start by saying this: I'm not going to sell a single cubic foot of gas to Austin or San Antonio. I'm not going to get into a squabble with a city. I'll sell you all the gas you want. If you want to sell some of it to them, that's your problem. But *you* are going to pay us."
>
> I bought it and sold it to Austin and San Antonio. And sure enough, they would always be late sending us a check . . .

Fortunately, said Soderberg, Houston Natural's lines tied into the Coastal States pipelines, which ran into LCRA plants. Thanks to Herring's political contacts, he was able to obtain gas from Houston Natural and some of the other companies. "He also was instrumental in getting customers together and negotiating new contracts," said Soderberg.

The "new contract" proposals created such dissension in the ranks of the LCRA's customers that the Pedernales and Bluebonnet coops felt the time was ripe for them to cancel their contracts to buy electricity from the LCRA. They felt they could save money by building their own "Texland" power plant. Herring offered some background:

> When I first went out to the LCRA, we had contracts with all our wholesale customers that we could never charge them more than nine mils per kilowatt hour. When this fuel thing came along, and gas went from twenty cents per million cubic feet to two dollars and fifty cents per MCF, that meant the fuel cost alone was more than nine mils.

One of the first things we had to do was try to get the customers to abrogate those contracts and sign new ones. I had forty-nine meetings with them. I finally worked out a contract and all of them signed it. But it was tough. Because they had to go home — a mayor, for instance — and say, "I signed a contract to raise your electricity bill."

But I kept after it. One of 'em came in and said, "I'll never sign a new contract. I've got the old one in a bank box and I've got the only key — and you'll never get it!" But I finally convinced them that we had only two choices: either we had to raise the rates or else we couldn't produce electricity because we'd go broke. They knew that all the time. And they finally bellied up.

Meanwhile, however, the LCRA appealed the Railroad Commission decision to the courts. Small's firm — which changed its name from Small, Herring, Craig, and Werkenthin to Small, Craig, and Werkenthin after Herring's appointment as general manager — did not participate in that case.

"Small's firm represented the LCRA almost exclusively for some time," said Hancock, "even in certain things that really weren't among their specialties. That was especially true while Charley was general manager."

Soderberg said he also used Small's firm most of the time after he became general manager. But he would also use other firms when a case involved expertise the Small firm didn't have.

When the LoVaca case began, objections were raised that Small would have a conflict of interests because he previously had represented Coastal. He denied any possible conflict but removed himself voluntarily from the case. The board then decided to hire Hume Cofer, who was in private practice in Austin and later became a district judge, to spearhead the LCRA's legal attack.

Lucksinger retired as controller and assistant general manager on December 31, 1974, but was retained to work with the lawyers on the Coastal States case until the trial ended in July 1975.

Cofer and Umstattd contended that Coastal and LoVaca had curtailed supplies to their contracted customers in order to sell gas at higher prices to new customers. They convinced a district court jury in Bryan to award the LCRA $25.2 million in damages — the difference between the contracted price of the fuel and what the LCRA actually had to pay for it.

The judgment was reversed by the Texas Supreme Court and

the case went back to the Railroad Commission, which began another series of hearings on the matter. With Small again representing the LCRA, Coastal States and LoVaca finally agreed on a settlement that was approved on December 26, 1977.

Under that agreement, the Coastal States gas-gathering system was given to a newly organized firm, the Valero Energy Corporation, and Coastal was required to put approximately $180 million into that firm's search for new gas. The Valero stock was held in trust for the LCRA and other customers who agreed to the settlement according to Small, one of the chief negotiators.

The LCRA's share of the settlement also included partial reimbursement of legal fees and the right to purchase at book value the lignite coal leases held by Coastal States in Bastrop, Washington, and Fayette counties.

Soderberg said the LCRA eventually received, over a period of four years, approximately $8 million from dividends and sales of the Valero stock by the trustee bank which held it. That money was refunded directly to LCRA customers, he said.

The problems with Coastal States prompted the LCRA to expedite its construction of generating plants with alternative fuel sources. In partnership with the City of Austin, the LCRA built two 600-megawatt, coal-fired generating stations near LaGrange — the Fayette Power Project. In 1974 it signed a long-term contract for western coal, to be hauled from Montana by railroad. But the days of cheap energy were over. The following year the State of Montana imposed a thirty percent severance tax on coal, and the railroad rates for hauling coal began rising rapidly.

Those factors later prompted a decision to build a third Fayette unit which would be powered by the lignite that was plentiful in the immediate area. The LCRA acquired lignite reserves estimated at 187 million tons — enough to fire a 400-megawatt plant for thirty years — on 17,000 acres within nine miles of the generating station. Although lignite is a much inferior fuel than western coal, the savings in taxes and transportation costs made it seem far superior.

But meanwhile, the LCRA paused on August 9, 1974, to dedicate its second thermal-generating plant — gas-fired, but with the capability for using fuel oil: the Thomas C. Ferguson Power Plant. It was located near Marble Falls, on the shores of Lake Lyndon B. Johnson (originally known as Lake Granite Shoals). Water from the lake would be used for cooling purposes in the $70 million, 430,000-kilowatt plant.

Furnishing gas for the Ferguson Plant was another matter which caused friction between the LCRA and Coastal States, and also, surprisingly, between Gideon and Lucksinger.

Small said Coastal's Oscar Wyatt took the position that he was obligated to supply gas for the plant but not to deliver it "off" his line.

"He said he would deliver it at Bastrop," said Small.

Lucksinger became angry with Gideon, he said, when the general manager contended that Coastal States needed a rate increase in order to build a pipeline from Bastrop to the Ferguson Plant.

"The only question," said Lucksinger, "was where the connection would be made with their lines. I finally suggested hooking up with them at Kyle and building the pipeline from there to the Ferguson Plant."

Lucksinger also objected strongly to Small's being retained to negotiate the contract with Wyatt since he previously had represented Coastal. But, he said, Gideon contended that did not matter since Small no longer was connected with Coastal.

Lucksinger had gone to Tulsa, Oklahoma, for the 1971 Christmas holidays and had suffered a heart attack while he was there. He was hospitalized for several weeks and could not return to Austin until about the middle of February in 1972. When he finally was able to return to work, he found that Small had worked out a contract with Coastal for the Kyle-to-Ferguson pipeline.

"Gideon and I were at each other's throats for a while," said Lucksinger, who felt the LCRA had been forced to contribute too much money for construction of the pipeline.

But personal animosities did nothing to spoil the dedication ceremonies for the Ferguson Plant on August 9, 1974. According to an article by Dave Allred in the August 15, 1974, issue of the *Marble Falls Highlander:*

> More than 300 people gathered in the hot sun for the 10 a.m. ceremonies and heard District Judge Herman Jones of Austin praise both Ferguson and the LCRA. He lauded Ferguson as a man of deep religious faith and as one who early recognized the potential of the Colorado River, and worked to achieve many improvements.
>
> Ferguson was named to the first LCRA Board of Directors

in 1935 by Atty. Gen. William McCraw. He served on the board on three other occasions, under appointments by Govs. Coke Stevenson, John Connally and Preston Smith . . .

Although the LCRA is a state agency, it receives no money from the Texas or federal governments, Judge Jones said. "You know it is unique when you realize that it is a governmental agency that lives within its own income."

Bill Petri of Austin, chairman of the LCRA's board of directors, presented Ferguson with a copy of the resolution naming the plant in his honor . . .

Herring presided over the hour-long ceremonies. He and Petri still were on good terms at that point — partly because the contracts for the Ferguson Plant had been let four years earlier, long before Herring became general manager. The plant had been designed by the notoriously non-union Brown & Root Company and built by the notoriously non-union H. B. Zachry Company of San Antonio. As a labor organizer, Petri had spent many years fighting both firms. He recalled:

> I caught hell on that Ferguson Plant. There were at least two or three hundred illegal aliens working up there — and this was while Gideon was still manager — and it was written into the contract that only citizens of this country could work on the project.
>
> Pat Zachry said at first that clause was illegal and he'd never agree to it. They told me that in Gideon's office and I told them to get Zachry on the phone. I said, "You tell him that I'll just enjoy going to court and hearing him say this clause is illegal." Schmitt called and told him: "Pat, the Board has just told me to tell you that that clause will be in there or you've got no contract. And the chairman is sitting right here."
>
> Zachry said he was going to sue us and I told Schmitt to tell him he could go ahead. The next day, Zachry called back and said he'd thought about it and he'd agree to that clause. But then what did he do? They'd run off about a hundred wetbacks one day and bring some more back in the next day. That went on all the time.
>
> Herring had told me that if he became general manager, he would assure me that Brown and Root and Zachry would not do any more work for the LCRA. I didn't even ask. He just volunteered that. I told him I'd appreciate it and knew the building trades would, that it was time they did some of the work.
>
> I think he volunteered that commitment because he thought

that's what I wanted to hear. "I'm not saying that he wasn't sincere — but that's sure not the way things worked out."

Herring insists that "Bill must have been dreaming." He said he never expressed any intention of giving any contract to anyone but the lowest bidder. Shortly after the Ferguson Plant dedication, he said, Brown and Root submitted the low bid for construction of the Fayette Power Plant, which was built on a 50–50 partnership basis with the City of Austin. Herring elaborated:

> The Bechtel Company, a union firm, designed the plant and wanted to submit a bid for construction on a cost-plus basis. There was no way we were going to let a cost-plus contract — and there was no way I was going to let the same firm that designed any plant have the contract for building it. I always wanted someone looking over the other guy's shoulder, instead of having one firm do everything.
>
> Bill's a labor man and I understand his position — but I don't agree with him, and I'm not mad at him. Actually, Brown and Root submitted a bid that was about twenty million dollars below the next lowest one. And Brown and Root actually paid its people as much as the union people were paid — and in some cases, even more. The non-union firms don't pad their personnel, such as having two people do something one person can do. So they can pay the same wages and still cut down on the costs. I didn't have anything against labor but I was looking after the LCRA.

According to the official minutes, the LCRA board considered on October 23, 1975, bids from four firms for construction of Units One and Two at the Fayette project. Brown & Root was low with a bid of $55,700,000, followed by the Austin Building Company, $73,095,000; H. B. Zachry, $84,620,000; and Mercury Construction, $95,713,000.

Chief Engineer Soderberg's written analysis of the bids noted that the Austin Building Company's proposal included six "exceptions and clarifications" to the specifications and that the bid submitted by Brown & Root was the most thorough and complete, with "no exceptions to the specifications listed in the proposal."

J. C. Strahan moved that the board adopt a resolution awarding the contract to Brown & Root. His motion was seconded by Hancock, then passed by a 9–2 vote with Petri and Bob Long voting against it. Voting "aye" were Voelkel, Dalchau, Dean, Hancock, Jungmichel, Cecil Long, Mayfield, Strahan, and Zercher. (Seymour was absent.)

Twenty years later, Herring was still proud of that decision's results.

> When we began making those plans in 1974, we estimated the total cost of the project would be about five hundred million dollars and the first unit would go into operation no later than the end of June, 1979. The first unit went into operation June 1, 1979. Unit Two came on line eleven months later, also a couple of months ahead of schedule. And the total cost was four hundred and fifty-four million dollars. We finished the project early and under budget — something no utility in the nation had done for ten years.

The Brown & Root bid was for labor only; the LCRA furnished all materials except for a few minor, mostly disposable items. Petri contends that Brown & Root submitted an unrealistically low bid on the Fayette Project but then swelled its profit by successfully submitting "change orders" which added to the cost. Herring and Soderberg both insist that is not true. Said Soderberg:

> We told Brown and Root at the start that there would not be any change orders unless we approved them in advance. The job was to be done according to the plans and specifications.
>
> They were one of the few firms that would give us a definite price. Bechtel normally did "turn-key" jobs — handling the engineering and construction on a cost-reimbursable basis. We always took the position that this was like the fox watching the henhouse. If you have one firm designing it and another firm doing construction, you've got a watchdog to make sure they do it right. And the City of Austin, our partner in the project, was adamant about the fact that their charter would not allow them to award a contract without taking bids.

One major change order, which added $7,627,855 to the cost of the project, was approved by the board on June 16, 1978, according to the official minutes. In a letter to the board, Soderberg said the additional expense was due to a delay of seven months and thirteen days by the LCRA in delivering structural steel to the job site for construction of the Unit Two boiler.

Soderberg said the LCRA was obligated to deliver the steel by July 1, 1977, but did not do so until February 10, 1978. The delay, he explained, was caused by an extensive search for the best type of fuel to use — which would affect the design of the boiler. That search began in July of 1975 and included evaluation of fifteen proposals, he said. It ended on April 14, 1977, with a recommendation for a fifteen-year contract with Atlantic Richfield to supply coal for Unit Two. Soderberg's letter continued:

> Even though this delay in obtaining a coal supplier was the main cause of the delay in starting Unit Two boiler design, it did result in a considerable savings to our customers. The amount saved will amount to approximately $160 million when compared to the Decker coal (with the Montana tax included) over a 15-year period, or a minimum of $32 million even if the tax is repealed.
>
> Brown & Root has estimated the added cost of extending the Unit Two schedule by eight months to be $6,043,445. This would establish a new trial operation date of Nov. 1, 1980, and a commercial operation date of Feb. 1, 1981.
>
> As an alternate, Brown & Root, Inc. submitted a proposed lump sum price adjustment of $7,627,855 to accelerate the construction effort and maintain the original March 1, 1980, trial operation date for Unit Two.
>
> The differential fuel cost between June 1, 1980, and Feb. 1, 1981, commercial operation, has been determined to be $13,945,000 based on $2.50 gas vs. $1.85 ARCO coal, and a 65% capacity factor . . .
>
> In view of the fuel differential, the most economical solution is to maintain the original schedule and pay for the accelerated construction effort.

That proposal was approved by the board on a vote of 11–2 with Hancock abstaining. He explained that a company in which he had an interest was negotiating a possible contract with Brown & Root and he wanted to avoid any possible appearance of a conflict of interest.

Petri and Zercher cast the "no" votes. In favor were Dean, Strahan, Dalchau, Jungmichel, Cecil Long, Mayfield, Charles Schreiner III, Seymour, Harry Shapiro, J. R. Thornton, and Milton Y. Tate, Jr.

Herring and Soderberg said that after construction of the plant had been completed, Brown & Root submitted a bill for several million dollars in "extra charges."

Herring said he told Ben Powell, Jr., who was the general counsel for Brown & Root, that there were not any "extras" and that the total price had been agreed upon in the contract.

"They threatened to sue us and I told 'em to go ahead and

sue," said Herring. "They finally wrote us a letter saying they'd waive the extra charges."

On July 27, 1979, an overflow crowd of 400, forced inside by rain, attended dedication ceremonies at which Fayette's first two units were named in honor of Sam K. Seymour, Jr. Congressman Pickle, Mayor Carole McClellan of Austin, and Herring were among the dignitaries who paid tribute to Seymour, then eighty-three years old, for his longtime service on the LCRA board.

Seymour and his wife, nicknamed "Muddie," received a standing ovation as she unveiled a plaque naming the two units for him. The inscription on the plaque read: "In tribute to his distinguished record of accomplishments and dedicated service to the people of Texas, these two coal-fired generating units are named for Sam K. Seymour, Jr., who has given generously of his time and energy for 34 years to the achievements of the Authority."

Pickle praised the LCRA for deciding to use coal at the plant and said: "Our brethren in the north and midwest should thank the leaders of Austin and the LCRA for freeing up more natural gas for their residential and industrial use."

But he also urged the LCRA and other Texas utilities to consider developing the use of "a home-grown fuel called lignite."

By the time Herring left the LCRA in August of 1981, the Authority had acquired vast lignite reserves in its move to diversify fuel sources.

In a 1981 interview with John Williams for the LCRA *River Review,* Herring was asked to summarize his major accomplishments as general manager. He cited the transition to alternative fuels, the successful construction of the Fayette Power Plant, the LCRA's excellent financial reputation (which made its bonds easy to sell) and the prosecution of Coastal States for breaking its contract. He also mentioned the revision of flood control operations for Mansfield Dam and Lake Travis, after two years of negotiations with the U.S. Army Corps of Engineers. Herring went on:

> The new regulations give LCRA more flexibility in providing equitable protection to residents above and below the dam, with a minimum of damage and wasted water. We have as good a record on that score as anyone who tries to harness a river.
>
> Over the years, we've made steady improvements in our electric system, and today we have what many consider to be the best operations control center in Texas, which also serves as the control center for the South Texas Interconnected System. That's

a power-exchange network set up so that the area utilities can help each other during emergencies — though the need for extra capacity by some utilities had prompted daily power sales between various electric companies. When that happens, the technical arrangements are usually set up in our control center.

I have a feeling, too, that over the years we have built an excellent reputation with the private utilities of Texas. There used to be some feeling of animosity between public power and private power. I think that has dissipated and the two work together — we help them as much as they help us — and the old antagonism, if that's what it was, is gone. I think that's something that didn't just happen — it's a result of the reputation the Authority has built for itself . . .

There's been tremendous demand for our electricity and water, and we will have to contend with even greater demands for those services. Although we've had an excellent environmental record in protecting the Colorado's water quality and by building what the Environmental Protection Agency considers to be the cleanest coal-fired power plant in this region, the growing demand for our resources will require LCRA to be as vigilant as ever in helping keep our water and air and land as clean as possible.

There are quite a few people who do not like the growth and the changes they've seen in the last few years, and these people would prefer not to let any more people in, not to plan for new developments or new industries. I can understand this feeling, to some extent, but these people forget that growth and change can be beneficial.

The building of the LCRA dams and the impounding of the Highland Lakes was one of the most far-reaching, even traumatic, changes in the Central Texas landscape and environment. And yet I've heard no cries that the dams be torn down to return the river valley to its natural state. We've all come to take the benefits of flood control, electricity, abundant water, scenery and recreation for granted . . .

It's been a long, hard eight years, but the rewards of seeing what you can do is worth all the work it takes to do it. We — the employees and officers of the Lower Colorado River Authority — are a family of people that are trying to do a job. And if anyone wants to criticize or take on the LCRA, they're taking on the whole family and we close ranks.

We never start out to lose — and we seldom do.

21

The Peace Maker

Herring told the board on January 2, 1981, that he wanted to retire as soon as a successor could be found for the general manager's job. Board Chairman Harry Shapiro of San Saba headed a selection committee which conducted an extensive search for a suitable replacement. It included Tom Dean, a rancher from Cherokee; John M. Scanlan, an Austin lawyer; and two businessmen, Marvin Selig of Seguin and E. B. (Tex) Mayer of La Grange.

After the committee had spent six months screening candidates without success, Herring grew impatient and told the board he wanted to leave in August. Meanwhile, Soderberg had informed the board that he was not interested in the job; but he was there and, obviously, he knew the ropes.

Tom Dean, a good friend of Soderberg's, came into his office one day and said the board had been unable to find anyone the members would feel comfortable with as general manager and that Herring was wanting to get away. "He asked if I would consider taking the job on an interim basis while they continued looking," Soderberg remembered. "I told him I would think about it."

He wanted to discuss the matter with his wife, Mary Virginia.

She was concerned that it might cause me some health problems. I was in excellent health but she knew all the things that had been going on and she knew my nature. She knew I would worry a lot in that job. She knew I was a very sensitive person and that the in-fighting among board members and among the politicians might worry me to the extent that some health problems would develop.

But she also said that if they were going to do it on an interim basis and would continue to search for a permanent general manager, maybe I ought to try it and see how I got along. "Otherwise, you would always wonder whether you should have tried it," she said. "That would always be in the back of your mind. If you feel comfortable with it, you may even want to apply for it."

Soderberg was reminded, during the interview several months after he retired, that everyone seemed to think of him as an easygoing, "laid-back" sort of person, and that he must have kept up a pretty good front to keep them from knowing how much he worried.

"I always felt like you get a lot more done by trying to work *with* people than by pounding on the table and making demands . . . I think I was successful in that. And I never did have any health problems," he said.

His relations with the board were excellent, in his opinion. When he was appointed, the managers and employees were told to disregard the word "interim" in front of his title. "They said I was the general manager, so far as the board was concerned, and told me I should not be reluctant to do anything that I wanted to do for the benefit of the LCRA," he said.

Soderberg's first day in the top post, August 17, 1981, was not nearly as eventful as his first day as chief engineer had been.

It was pretty calm and really kind of a happy time. I had a staff meeting once a month. I invited the managers from all the districts to come in and have a little chit-chat, to tell us what they had been doing during the previous month and what was going on. Then I would tell them about the board's actions and what it was thinking and we'd discuss things.

That first meeting after I became general manager, I walked upstairs to the Board Room and all these people were standing around. They started playing "Hail to the Chief" and they had a big sign on the wall that said, "YOU'VE COME A LONG WAY, BABY — FROM MAIL CLERK TO GENERAL MANAGER."

He had indeed. And like most longtime employees of any organization, he had noticed a lot of things along the way which he thought ought to be changed. He plunged enthusiastically into his new duties.

"We had never had a data processing division on a company-wide basis . . . So I set up a company-wide data processing center and hired someone to supervise it. I set up a water resources department and a public affairs department — and I started trying to establish some new parks," he proudly reflected.

Soderberg said the board's reluctance to go along with Petri's park plans was due mainly to the idea that LCRA funds should be used to develop the recreational facilities.

> The board was somewhat reluctant to take funds from the electric customers and use them for parks instead of using them to keep the rates down. When I got in, I started trying to get other entities, such as counties and the State Parks and Wildlife Department, to go into partnership with us — where we would furnish the land and control what went into the parks, but they would develop them.
>
> We joined hands with Travis County, for instance, to develop several parks. Pace Bend was one of the first ones, I guess — and it's the largest, with 1,520 acres. We had a parks department that would approve the layouts and how the parks would be developed. That way, we had no LCRA funds in the parks — just the land. And we had some control over what went into them. That worked out very well and we established a number of parks with county cooperation.

Meanwhile, however, Soderberg also inherited a number of knotty problems — some of long standing and some involving lawsuits. And, for a while, he seemed to get more help than he really needed.

The board had been accustomed to strong-handed leadership from Gideon and Herring. Soderberg said that with Petri there, the board acted as a rubber stamp. "But then the pendulum swung the other way — and it swung too far," he declared. "Once Herring announced he was leaving, the board members started getting more and more involved in the operations, mainly through their committees. They continued to do that while I was interim general manager."

Hancock, a former chairman of the board, agrees with that assessment.

> You have to remember that Elof was an *interim* general manager while a search went on. That's when we established this committee system to give some guidance to the general manager and the staff. I think we went too far and became too active.
>
> The search committee, of course, never did come up with anyone strong enough to recommend. So we looked around and said, "What's wrong with Elof? He knows the system from one end to the other and all the employees like him. And we all get along fine with him." The answer was that there wasn't anything wrong with Elof—unless it was that he wasn't mean enough. He was too nice a fellow, very likeable—I think that was the general appraisal of him.

It was about the first of October, Soderberg recalls, when Dean once again visited his office and asked him if he would consider taking the job on a permanent basis.

"The employees seemed to be ecstatic that I was general manager. I was one of the boys who had grown up with the company and they were all very cooperative with me," he said. After discussing the situation with his wife again, he told the board he would like to be considered.

On November 19, 1981, Shapiro presented to the board a resolution which had been approved by the Selection Committee and it was adopted unanimously. It noted that the committee had spent ten months screening more than 200 applicants from throughout the nation and had interviewed twenty of them while Soderberg was performing the general manager's duties. It said that Soderberg, during his thirty-nine-year career with the Authority, had developed an understanding and knowledge of its responsibilities and had acquired the necessary qualifications for a general manager. It added:

> WHEREAS in his capacity as Chief Engineer, Mr. Soderberg has substantially contributed to the Authority's modern-day successes, including the completion of the coal-fired Fayette Power Project within its original cost and time schedule, the assurance of reliable power and energy and low electric rates, and the preservation and efficiency of the state's emergency power-exchange network; and
>
> WHEREAS, individual members of the Board concur that

Mr. Soderberg has performed excellently during this interim period, not only as Acting General Manager but also as Chief Engineer, in keeping the Authority operating smoothly, reliably and efficiently; and

WHEREAS Mr. Soderberg has, over the years, earned the support, respect and admiration of utility executives throughout the nation, representatives of state and federal agencies, the Authority's suppliers, and the unanimous endorsement of the Authority's wholesale-customer organization and most especially the support of its employees; and

WHEREAS it is the determination of both the Selection Committee and the Personnel, Compensation, Pension Trust and Benefit Committee that the knowledge, experience, reputation and performance of Mr. Soderberg exceeds not only the minimum standards that have been set for consideration of a General Manager, but also the qualifications of the other candidates that have been considered for this position,

NOW THEREFORE, BE IT RESOLVED by the Board of Directors of the Lower Colorado River Authority that, effective immediately, the Authority's General Manager shall be Mr. Elof Soderberg; that this Board wish him well and pledge its support in his difficult task of leading LCRA to become a more efficient, effective and meaningful organization in accordance with its objectives and the enabling legislation that created LCRA . . .

Almost immediately, Soderberg began implementing reorganizational plans which were designed to disperse authority more widely and bring more key employees into "officer" status.

One of the first things he did was contact his old friend R. L. Hancock, then director of electric utilities for the City of Austin, and offer him the newly created post of executive director, engineering and operations. That superseded the chief engineer's position, which Soderberg had continued to handle while serving as interim general manager.

Hancock had graduated from Austin High School in 1943 and Soderberg in 1944, but they first became friends while attending The University of Texas. They had worked closely together on the Fayette Power Project, especially on the concept of making it a joint facility for the LCRA and the City of Austin.

Soderberg also hired John D. Heaton of Boston, an officer of a financial holding company there, as executive director for finance and administration.

Hancock said that Soderberg made changes to allow for more participatory management and more aggressive development in water quality. He cited Bill West as an example. West had been working with both electricity and water as manager of system operations before Soderberg made him director of water operations "so he could devote all his time to water matters."

In 1982 the LCRA established a water-quality monitoring program. It began taking water samples on a regular basis at forty sites along the river to evaluate quality and establish definitive water standards for the future. (The Authority had been regulating, licensing, and inspecting septic tanks along the river since 1971 as a part of its pollution control efforts.)

Board member John Hancock (no relation to R. L. Hancock) felt the LCRA took a highly important water supply step on January 20, 1983, when it agreed to buy the Lakeside Irrigation Co., Inc., for approximately $5 million. That firm, headquartered in Eagle Lake, provided irrigation for 28,500 acres in Colorado and Wharton counties through 270 miles of irrigation canals.

The LCRA began operating Lakeside on February 1, 1983, as part of its responsibility to conserve and supply water for river-basin residents.

"If there is any future in farming," noted John Hancock, "we cannot afford to let such water rights fall into, or stay in, private hands."

The board also decided on January 20, 1983, to buy approximately 6,400 acres of lignite leases in Fayette County from Exxon Coal Resources, USA, for $10.2 million. Those leases, located in the Cummins Creek and Rabb Creek areas where the LCRA already owned leases on 18,000 acres (near the Fayette Project), hold lignite estimated conservatively as 70 million tons, Soderberg noted.

"The benefits to LCRA and its customers are significant," he declared at the time, "in that the Exxon leases increase our lignite reserves sufficiently to provide enough fuel to fire two 400-megawatt generating units."

Construction of a 400-megawatt, lignite-fired unit at the Fayette Power Project began in April of 1984. H. B. Zachry was awarded the general contract on a bid of $70,165,000, the lowest of six submitted. According to the LCRA's 1984 *Annual Report:*

This low bid, along with competitive markets for goods and services, a lower rise in the rate of inflation and timely planning by LCRA staff, enabled us to lower our cost estimate significantly for Fayette 3 and supporting facilities. The original estimate in 1980 — including the Fayette 3 unit, mining and fuel transportation facilities — was $1.3 billion. More recent estimates have lowered this figure to roughly $800 million — a significant decrease which enhances Fayette 3's economic advantages.

The new unit was scheduled to go into operation in 1988, using locally mined lignite. But large decreases in the prices of both coal and natural gas, along with reductions in railroad freight rates, prompted the LCRA to begin considering alternate fuel sources for Fayette 3.

Meanwhile, in another move aimed at improving water quality, the board on October 18, 1984, authorized studies of proposed regional water and sewer districts on both sides of Lake Travis and Lake Buchanan.

The action was taken at the urging of Robert M. (Bob) Tinstman, the LCRA director of administration and chairman of its water policy committee. Tinstman, a former city manager of Austin, said fast action was needed "to catch the runaway train of rapid growth" in the river basin.

"People are not waiting for us to see what we're going to do," he declared. "They're proceeding with their own development."

An article by Max Woodfin in the October 19, 1984, edition of the *Austin American-Statesman* noted:

> The studies will be the first major step toward getting the river authority into the business of providing water and sewer service, a move approved by the Board in June.
>
> Included in the information sought through the studies will be the type of water or wastewater plants to be built, the potential service areas, the cities or utility districts that could use the services, and the extent of LCRA involvement in designing, building and operating the facilities.

In keeping with his personality, Soderberg became known as a peacemaker during his tenure as general manager. He settled several old disputes but, in keeping with its own heritage, the LCRA continued to become embroiled in new controversies.

Soderberg's first real "armistice" agreement came in November of 1984. As Bill McCann described it in the November 16, 1984, issue of the *Austin American-Statesman:*

The Lower Colorado River Authority and two groups concerned about plans to strip-mine lignite in Bastrop County reached a settlement Thursday in a two-year battle over the mining issue.

The settlement calls for the river authority to relinquish mining leases on 3,700 acres just north of the city of Bastrop in return for opponents dropping a lawsuit over a wastewater discharge permit.

The settlement removes a major stumbling block to LCRA plans to begin operating a small strip mine on a 956-acre tract just north of the 3,700 acres. The settlement also virtually eliminates any possibility that the LCRA will develop a large mining project in the area, which is known as Powell Bend.

One of the concerns of area landowners has been that the river authority planned to use the small strip mine as a steppingstone for a major operation in the Powell Bend area. LCRA officials repeatedly denied they had such plans. But by giving up the leases, the river authority essentially kills any idea of a large mine there.

Plans call for workers to begin removing lignite next week at the small mine after completion of work on some mining facilities, LCRA officials said. An average of 200,000 tons of lignite is to be stripped yearly during a 10-year period.

The lignite, a low-quality coal, is to be shipped by rail 46 miles to the Fayette Power Project near La Grange. The lignite is to be mixed with higher quality coal from Montana and Wyoming and burned at the two Fayette units, owned jointly by the river authority and the City of Austin.

Burning the lignite at the plants will save the river authority and its customers about $5,000 a day in fuel costs, said LCRA General Manager Elof Soderberg . . .

Even while this lignite skirmish was being put to rest, the long-smoldering controversy between the LCRA and the Colorado River Municipal Water District (CRMWD) over the proposed construction of Stacy Dam was flaring anew. The CRMWD wanted to build that project near Ballinger, about fifty miles east of San Angelo, just below the confluence of the Colorado and Concho rivers. Proponents said that reservoir could provide a dependable water supply for such West Texas cities as Odessa, Midland, Big Spring, Snyder, San Angelo, and Midland. But the LCRA had argued that it would cut off water from reaching the Highland Lakes and make it impossible to meet downstream water needs in the future.

The LCRA also contended that the reservoir would not be economical or practical because its evaporation rate would be so high that the CRMWD would lose more water through evaporation than it would actually gain in storage.

The Texas Water Commission granted a permit on April 10, 1979, for construction of the dam and reservoir. After a long, bitter court fight, the LCRA apparently won the battle on November 14, 1984, when the Texas Supreme Court reversed the commission's decision.

But the war was not over. The next day. Governor Mark White told reporters the Supreme Court's ruling jeopardized the latest in a long line of legislative efforts to develop a statewide water plan. Attorneys for the Stacy proponents filed a motion for rehearing and on January 9, 1985, the Supreme Court withdrew its decision, remanding the case back to the Texas Water Commission.

When several key West Texas legislators announced that they saw no reason to support a proposed $600 million statewide water plan which did not include the Stacy reservoir, Governor White really went to bat for the project. He met on January 21 with the LCRA Board of Directors, Soderberg, LCRA General Counsel John Bagalay, and some of the heaviest hitters in the legislature. They included Speaker Gib Lewis, Senator Grant Jones of Abilene, Senator Bill Sims of San Angelo, Representative Tom Craddick of Midland, an aide to Lieutenant Governor Bill Hobby, and Senator John Montford of Lubbock, Senate sponsor of the key water bills.

LCRA Board Chairman John W. Jones of Brady emerged from the meeting in the governor's office hopeful that a solution to the problem could be reached.

"The Governor and Speaker Lewis and the legislators all reiterated the importance of the statewide water plan," said Jones, "and expressed their feelings that the LCRA and the Colorado River Municipal Water District should negotiate a settlement."

Soderberg and Owen Ivie, general manager of the CRMWD, actually had been trying to do just that for several months, but without any publicity.

"We were anxious to come up with a settlement," said Soderberg. "The dispute was a thorn in everybody's side . . . I always felt that reasonable people can sit down and resolve their differences without taking legal action."

They finally took a clue from an agreement reached years earlier between the LCRA and the CRMWD on the Robert Lee Dam, which created the E. V. Spence Reservoir on the Colorado in Coke County. That matter, somewhat similar to the Stacy problem, was settled while Gideon was general manager — and with Herring representing the LCRA as its legal counsel.

A very comprehensive water release plan was finalized for Stacy. Soderberg explained that when Travis and Buchanan are below certain levels and the Stacy reservoir is above a certain level, water had to be released to LCRA. "That protects our customers from being deprived of water," he said. "The agreement was patterned after the one on Robert Lee but is much more comprehensive and detailed."

He also said the new reservoir would be filled in stages and that it would still have an effect on downstream Colorado River water users, but that would be spread over many years instead of hitting suddenly.

Governor White, Lieutenant Governor Hobby, and Speaker Lewis all continued to push for an agreement, which riled at least one of the LCRA board members. The board gave preliminary approval to the agreement by an 8–2 vote on January 29, 1985, with Scanlan and Jay Anderson, a rice farmer from Eagle Lake, voting against it.

Anderson later complained about the pressure from state officials and legislative proponents of the statewide water plan.

"We were being asked on the altar of a statewide water plan to give up something the Supreme Court had already given us," he declared. "We're being asked to 'play God' with the lives of our various constituencies."

The LCRA board had turned down several earlier proposals for a truce on Stacy. Jones, as chairman, was instrumental in winning acceptance of this plan. But final approval was still two weeks away when the board, Governor White, and a host of other dignitaries paused on February 8, 1985, to celebrate officially the LCRA's fiftieth birthday.

While open houses were being held at twenty-three LCRA locations throughout the agency's service area, White delivered the keynote address at a ceremony in the rotunda of the State Capitol.

"The LCRA ushered in a whole new way of life — modern life — by harnessing the energies of the Colorado," he declared.

White also commended the LCRA for negotiating the Stacy

settlement the week before. He noted that it offered a graphic example of the fact that political compromise was still alive, just as it was in 1934 when Governor "Ma" Ferguson forged a compromise with the legislature to create the Authority.

Cactus Pryor, an Austin radio entertainer who served as master of ceremonies, responded to that statement by declaring: "You might say that Old Man River was saved by old Ma Ferguson."

Lady Bird Johnson represented her late husband, who had placed such an important role in the LCRA's development throughout his political career. And two members of the original LCRA board were on hand: former U.S. Senator Ralph Yarborough and Judge Thomas Ferguson.

"As a young man," said Ferguson, "I was filled with many of the dreams and hopes of those who hoped that the Colorado would be harnessed and directed to useful purposes. Those dreams have not only been met but far, far exceeded."

Herring praised "the people climbing the poles, designing the plants, going out at midnight in freezing weather . . . who have made the LCRA a tremendous success. And you'll find that loyalty, that understanding and that closeness among all the 1,500 or more employees of the LCRA."

Also participating in the program were Board Chairman Jones, former Chairman Sam Seymour, E. "Babe" Smith, Soderberg, Gideon, and Congressman Pickle.

While the celebration capped a half century of progress, it also marked the beginning of one of the LCRA's most momentous years.

On February 21 the board gave final approval to the Stacy plan by a 10–0 vote, with Scanlan and Anderson both absent. But Soderberg and Bagalay, the LCRA's general counsel, said the issues involved were "so important and so sensitive" they could not be revealed until the agreement had been approved by the CRMWD.

That approval came quickly. Ivie and other proponents were delighted to wind up a struggle with the LCRA that had started about twenty years earlier; indeed, one which could trace its origins all the way back to the creation of the Authority.

Although they received a permit to build the dam, their fight was far from over. They still had to contend with such obstacles as one raised by environmentalists who claimed the construction would destroy the habitat of the "Concho Water Snake" and perhaps make it extinct.

But so far as Soderberg and the LCRA were concerned, at least one more long and bitter feud had become extinct.

Several weeks later, the legislature finally passed the statewide water plan.

22

A New Beginning

On February 21, 1985, the LCRA and the City of Austin filed suit against the Decker Coal Company of Montana, seeking to cancel their twenty-five-year contract with the firm. The suit charged that Decker's prices for coal — up to $30 a ton — were "unconscionable" and sought damages for prior overcharges. The LCRA and the City of Austin both announced they would stop taking deliveries of Decker coal on April 21.

Decker promptly filed a countersuit, claiming that the firm could suffer at least $1 billion in damages if the LCRA and the city did not comply with their contract to take two million tons of low-sulfur coal a year.

The filing of the suits climaxed a remarkable year in the power-generation business. As noted in the LCRA'S 1985 *Annual Report:*

> For the first time since the Energy Crisis in the mid-1970s, the price of natural gas, western coal and rail transportation fell sharply during the 1985 fiscal year — the result of market conditions, increased competition among suppliers, and negotiation by LCRA of new contracts for buying and shipping fuel to its power plants. Fuel costs per million BTUs dropped by roughly a third. "That's a significant cost saving for LCRA, as traditionally al-

224

most 80% of our operations and maintenance expenses are for
fuel purchases," says R. L. Hancock.

The prices continued to drop for some time. In April the
LCRA and the city began buying coal from other firms for around
$5 a ton. But knowing that the costs probably would start rising
again eventually, they continued to nurture the lignite concept as a
viable alternative.

And, meanwhile, lignite had begun playing a more important
role in the Texland controversy — the long fight waged against the
LCRA by the Brenham-based Bluebonnet and the Johnson City-
based Pedernales Electric coops. Angered by being "forced" to sign
new, higher-cost contracts with the LCRA in 1974 as a result of the
Coastal State gas price increases, the two coops decided they could
save money by leaving the LCRA and building their own generating plant.

The two coops originally planned to build three 500-mega-
watt, coal-fired plants near Rockdale (in Milam County) with the
help of the LCRA. When the LCRA decided it was more practical
for it to expand the Fayette Plant with a third unit, the two coops
agreed to build the Rockdale project on their own. But they filed an
application with the Texas Public Utility Commission in 1981,
seeking a permit, and it was turned down.

Fred Werkenthin, who was representing the LCRA in opposing the application, declared that the PUC was correct in ruling
that the project was not needed.

"The Commission said the Texland project lacked financing,"
he noted, "and that the LCRA was handling its business affairs
properly."

The Texland proponents then took the matter to court. In November of 1983, District Judge Charles Mathews ordered the case
sent back to the PUC, declaring that the standards which it had applied to the feasibility of the proposal were too stringent. The PUC
joined the LCRA in appealing that decision and the Third Court of
Appeals in Austin ruled on September 11, 1985, that it should be
reconsidered by Judge Mathews.

But meanwhile, the two coops had devised another plan. Since
municipalities were exempt from PUC regulation, Pedernales and
Bluebonnet decided to join hands with Johnson City to build a $3
billion plant on the same Milam County site. Johnson City, a town

with a population of slightly more than 900, proposed to issue $1.6 billion in revenue notes to finance its share of the project.

A court suit on that proposal also was pending as Soderberg went quietly about the business of trying to negotiate a settlement, which he had said was one of his main goals as general manager.

The breakthrough came on September 16, 1985, just five days after the Court of Appeals ruling was announced. The two coops, in separate actions, accepted a settlement offered by the LCRA and a final agreement was executed on September 20, 1985.

Essentially, it provided that (1) all parties involved in the Texland/Johnson City matter would dismiss or drop all lawsuits, counterclaims, and appeals; (2) the LCRA would drop its opposition to the proposed Rockdale Power Project; (3) the two coops would be released from their power contracts with the LCRA *if* the Johnson City coalition could secure the necessary financing, fuel sources, and permits to get construction under way by December 31, 1987; and (4) should such progress be made, the two coops would be permitted to gradually phase out their power supplies from the LCRA but, if not, they would be permitted to sign new, long-term contracts with the Authority.

Soderberg said the agreement provided adequate protection for LCRA customers because, if the proposed plant should be built, the LCRA would not have to construct new facilities in order to keep pace with the anticipated growth in power demands. In effect, he said, it probably would preclude the necessity of building a fourth unit at Fayette. Also, he added, the LCRA would continue to furnish electricity to the coops through June 30, 1999, but any growth in their needs would have to be taken care of by the new plant.

Soderberg still expressed doubts that the project could be financed and built.

"But whether it is or not," he said, "I think our customers are protected."

During the same meeting at which it approved the "Texland" settlement, the LCRA board turned thumbs down on a long-sought, $234 million dam and reservoir between Columbus and La Grange — by a vote of only 7 to 5. Technically, the vote came on the U.S. Bureau of Reclamation's request for board endorsement of its plans to seek $600,000 from Congress for advanced studies on the project.

The proposed reservoir, which would have flooded about 15,000 acres in Fayette and Colorado counties, had drawn strong opposition from local landowners and from public officials in Fayette County. The dam would have been built at Shaws Bend, four and one-half miles upstream from Columbus. The board expressed a willingness to consider other sites, but that seemed unlikely after the ten-year search to find the chosen location.

Board Member Patricia "Pinky" Wilson of La Grange said Fayette County already had surrendered much of its land to the LCRA for lignite mining operations.

"If the LCRA continues at the rate they are going," she declared, "there won't be a Fayette County. It will be LCRA County."

Jack Miller, a member from San Saba, said flatly: "If the people down there don't want it, I don't believe in forcing it on them."

Voting against the proposal, in addition to Wilson and Miller, were Jack Littlejohn of Fayette County, Ken Dixon of Llano County, Charles Matus of Johnson City, Scanlan, and Hancock.

Those voting for the project were Burton LeTulle of Bay City, Marvin Selig of Seguin, Martin McLean of Marble Falls, Merritt Schumann of New Braunfels, and Jack Martin of Austin.

Jay Anderson of East Bernard was absent and Cecil Long of Bastrop abstained. Jones, the chairman, would have voted only in case of a tie.

Soderberg had announced on August 22, 1985, that he planned to retire the next April. His forty-three-year career with LCRA, and especially his accomplishments as general manager, drew high praise from Board Chairman Jones.

"Having worked his way up from mail clerk," said Jones, "he knows the LCRA better than anyone and will be sorely missed. It will be a challenge to find a qualified individual who can measure up to the integrity, loyalty and dedication demonstrated by Elof Soderberg."

Jones named five members of the board to serve as a selection committee and recommend a successor to Soderberg.

Unfortunately, Soderberg's last days as general manager were marred, to some extent, by accusations of misconduct, conflicts of interest, theft of equipment, and favoritism in choosing a contractor on the part of a few of the Authority's more than 1,600 employees. The allegations, none of which were aimed at Soderberg, included charges that some LCRA employees had taken hunting and fishing trips paid for by LCRA contractors, had used house trailers at construction sites for parties that included strip poker games and sexual activities, and had used LCRA equipment and workers for personal business.

After a five-month investigation by Daniel Hedges, a former U.S. attorney hired by the LCRA to conduct the probe, Soderberg's successor forced two employees to resign and took disciplinary action against six others. The full report was sent to the Travis County district attorney, who apparently found nothing worth prosecuting in it.

A highly perceptive article by David Matustik and Bill McCann in the *Austin American-Statesman* on October 7, 1985, commented on the search for a new general manager:

> While the Lower Colorado River Authority is under the closest scrutiny in its 50-year history, the search is on for a leader to guide the agency into new arenas.
>
> A committee is looking for a general manager who has management skills to direct a staff of 1,600, and personality and savvy to deal with a 15-member board of directors who reflect the diverse and sometimes conflicting interests of the Authority's 10-county region.
>
> The manager will have to balance the internal business of the organization with a new priority of external public relations.
>
> What the board would like is a mixture of the good traits of the past two managers — the dependability and integrity of retiring manager Elof Soderberg and the political knowledge and skill of his predecessor, Charles Herring.
>
> What they do not want is Soderberg's soft-spoken style, which has allowed some staff officials to run things their way at times, according to board members. Nor do they want such a dominant personality as Herring, whose aggressiveness and sometimes unyielding stances on issues meant no compromise with the board or the public.
>
> Merritt Schumann, a member of the committee, said the panel believes the general manager "is a person who can be a good communicator" for the agency . . .
>
> Board members said each (Herring and Soderberg) was right for his era — but not necessarily right for today.
>
> Herring used political instinct to defend the agency when the legislature discussed its scope. Herring was comfortable in the game of power-brokering.
>
> Soderberg does not have Herring's ability to mesmerize a

crowd or his liking for a confrontation. Soderberg, the engineer, guided the river authority into the development of lignite-fueled power generating plants and lignite mining operations . . .

The LCRA board found its man in November of 1985: S. David Freeman, a former chairman of the Tennessee Valley Authority who also had served as an energy advisor to Presidents Lyndon Johnson, Richard Nixon, and Jimmy Carter.

LCRA Board Chairman Jones announced Freeman's selection on November 21, declaring: "Mr. Freeman's credentials are impeccable. The LCRA Board feels he has the necessary experience and capabilities to ensure that LCRA will continue to serve its customers, effectively and efficiently, into the next century. He will have the advantage of Mr. Soderberg's assistance and counsel until Mr. Soderberg's retirement is effective next April 6."

Freeman, who received a bachelor's degree from Georgia Institute of Technology in 1948 and a law degree from the University of Tennessee (where he graduated first in his class) in 1956, thus became the sixth general manager in the LCRA's history.

After being selected from 206 persons who were considered, Freeman described himself as a "participatory" manager:

> I'm going to get out with the folks. My management style is not to sit behind a desk and issue orders. It's to get the folks involved in the decision together and sit down and lock the door and throw away the key until we reach a decision.
>
> I love for people to argue with me. If there's anybody that thinks I want them to try and guess what I think, they're wrong. I like to believe that the best idea can survive and will be recognized by the whole management team.
>
> As part of our goals, we will increase our emphasis on conservation and environmental programs. I think it's encouraging that very few of these programs are completely new to LCRA; rather, they expand on well-established groundwork.

Freeman joined the Authority in January of 1986. Soderberg's retirement the following April marked the end of an era for the LCRA as it began its second half century under Freeman's leadership.

Regardless of what happens during the next fifty years — or even the next one hundred years — it seems unlikely that anyone connected with the LCRA will ever take lightly these words written by Lyndon B. Johnson in 1958:

Of all endeavors on which I have worked in public life, I am proudest of the accomplishment in developing the Colorado River.

It is not the damming of the streams or the harnessing of the floods in which I take pride, but rather in the ending of the waste of the region.

The region — so unproductive in my youth — is now a vital part of the national economy and potential. More important, the wastage of human resources in the whole region has been reduced. Men and women have been released from the waste of drudgery and toil against the unyielding rocks of the Texas hills. This is true fulfillment of the true responsibility of government.

Appendix A

Specifications of Dams

Lake Buchanan / Buchanan Dam

The structure was originally named Hamilton Dam by an Insull utility subsidiary that began the project in 1931. Construction was suspended in 1932 when Insull bankrupted. LCRA completed construction 1935–37. Both the dam and lake are named for U.S. Representative J. P. Buchanan, who helped secure federal funds for completion of the dam.

Built primarily to supply hydroelectricity and water storage, the dam is a series of arches, considered to be the longest multiple-arch dam in the nation. It contains thirty-seven floodgates.

THE DAM is 145.5 feet high. The length is 10,987.55 feet. At the base, it is 215.11 feet thick; at the top, it is 34 feet thick.

THE POWER PLANT'S three units provide a generating capacity of 36,000 kilowatts.

THE RESERVOIR is 30.65 miles long, and its maximum width is 4.92 miles. The lake covers 23,060 acres, and its capacity is 992,475 acre-feet. All figures are for the reservoir filled to the top of the conservation and power pool — elevation 1,020 feet above mean sea level (msl).

Lake Inks / Inks Dam

Constructed 1936–38, the structure was originally called Arnold Dam. Both dam and lake are named for Roy B. Inks, a member of the first LCRA board of directors. Built for hydropower, Inks Dam has no floodgates, and the power plant is the smallest in the chain.

THE DAM is 96.5 feet high. The length is 1,547.5 feet. At the base, it is 76.1 feet thick; at the top, it is 16.5 feet thick.

THE POWER PLANT'S one unit has a generating capacity of 12,000 kilowatts.

THE RESERVOIR is 4.2 miles long, and its maximum width is 3,000 feet. The lake covers 803 acres, and its capacity is 17,545 acre feet. All figures are for the reservoir filled to the top of the power pool — elevation 888 feet msl.

Lake LBJ / Wirtz Dam

Both dam and lake were originally called Granite Shoals. Constructed 1949–50, the dam was renamed in 1952 for Alvin J. Wirtz, the first general counsel and "father" of LCRA. The lake was renamed in 1965 for President Lyndon B. Johnson, who substantially assisted LCRA as a U.S. representative and senator. Built primarily for hydropower, Wirtz Dam has ten floodgates but no spillway. Lake LBJ provides cooling water for LCRA's Thomas C. Ferguson Power Plant.

THE DAM is 118.3 feet high. The length is 5,491.4 feet. At the base, it is 80 feet thick; at the top, it is 12 feet thick.

THE POWER PLANT'S two units provide a generating capacity of 52,000 kilowatts.

THE RESERVOIR is 21.15 miles long, and its maximum width is 10,800 feet. The lake covers 6,375 acres, and its capacity is 138,460 acre-feet. All figures are for the reservoir filled to the top of the power pool — elevation 825 feet msl.

Lake Marble Falls / Starcke Dam

Both lake and dam were originally named Marble Falls. Constructed 1949–51 for hydroelectricity, the dam was renamed in 1962 for Max Starcke, LCRA's second general manager from 1940 to 1955. The ten floodgates are designed in a "bear trap" fashion that fold down to pass through floodwaters. Both the lake and the dam are the smallest in the Highland Lakes chain.

THE DAM is 98.8 feet high. The length is 859.5 feet. At the base, it is 56.83 feet thick; at the top, it is 13 feet thick.

THE POWER PLANT'S two units provide a generating capacity of 32,000 kilowatts.

THE RESERVOIR is 5.75 miles long, and its maximum width is 1,080 feet. The lake covers 780 acres, and its capacity is 8,760 acre-feet. All figures are for the reservoir filled to the top of the power pool — elevation 738 feet msl.

Lake Travis / Mansfield Dam

Constructed 1937–41 by LCRA and the U.S. Bureau of Reclamation, Mansfield Dam serves as the primary flood-control structure for the lower river basin, as well as providing water storage and hydroelectricity. The dam was originally named after the nearby settlement of Marshall's Ford (and the U.S. Army Corps of Engineers still refers to the structure as Marshall Ford Dam). It was renamed in 1941 for U.S. Representative J. J. Mansfield, who assisted in development of the project. The dam contains twenty-four floodgates.

THE DAM is 266.41 feet high. The length is 7,098.39 feet. At the base, it is 213 feet thick; at the top, it is 30 feet thick.

THE POWER PLANT'S three units provide a generating capacity of 96,000 kilowatts.

THE RESERVOIR is 63.75 miles long, and its maximum width is 11,500 feet. The lake covers 18,929 acres, and its capacity is 1,170,752 acre-feet. All figures are for the reservoir filled to the top of the conservation and power pool — elevation 681.1 feet msl.

FLOOD CONTROL: Mansfield Dam is the only structure in the Highland Lakes chain specifically designed to contain floodwaters. When the elevation of the lake exceeds 681 feet msl, the water enters the flood pool (681–714 feet msl), and LCRA begins floodgate releases under the direction of the U.S. Army Corps of Engineers. The amount and duration of the releases will vary, depending upon weather and flood conditions above and below the dam.

Lake Austin / Tom Miller Dam

Named for an Austin mayor, Tom Miller Dam was built in 1938–40 atop the remains of two earlier structures built in 1890–93 and 1909–1912, each damaged by massive floods. The lake was originally called Lake McDonald; the two earlier dams were both called the Austin Dam. The dam was built to provide hydroelectricity and water. It contains nine floodgates.

THE DAM is 100.5 feet high. The length is 1,590 feet. At the base, it is 155 feet thick; at the top, it is 22.8 feet thick.

THE POWER PLANT'S two units provide a generating capacity of 14,000 kilowatts.

THE RESERVOIR is 20.25 miles long, and its maximum width is 1,300 feet. The lake covers 1,830 acres, and its capacity is 21,000 acre-feet. All figures are for the reservoir filled to the top of the power pool — elevation 492.8 feet msl.

Appendix B

LCRA Board of Directors

Eighty-three men and women have served as Lower Colorado River Authority directors since the agency's creation in 1934.

The original LCRA board in 1935 was composed of nine men, three each appointed by the governor, the attorney general, and the land commissioner. Appointees could come from throughout the state. Among the original directors, one lived in San Antonio, another in Abilene.

In 1941 the Texas legislature gave the governor sole authority in nominating directors. The legislature also changed the residency requirements for directors so that at least six had to reside within LCRA's ten-county statutory district.

The legislature has expanded the size of the board twice. In 1951 the board grew from nine members to twelve, giving each statutory county at least one representative on the board. In 1975 the legislature expanded the board to fifteen directors, to add at-large directors from the electric service area outside of the statutory counties.

Today the LCRA Board is composed of twelve directors representing the statutory district and three directors representing the electric service area outside the statutory district. Once nominated by the governor, the directors must be confirmed by the Texas Senate; terms are for six years. If the legislature is not in session, the appointees can immediately take their seats, pending Senate confirmation.

Of the twelve statutory-district directors, two are from Travis County and one each is from the nine other counties. These directors can be reappointed to successive terms. The twelfth director is appointed from one of the statutory counties other than Travis; however, this director cannot be reappointed, and his successor must come from a different statutory county. The three electric service-area directors cannot be reappointed, and their successors must also come from different counties.

The roster includes some famous names in Texas politics: Ralph Yarborough went on to become the state's attorney general, a district judge, and U.S. senator. John Connally later served as governor, as secretary of the Navy (under President John F. Kennedy), and as secretary of the treasury (under President Richard M. Nixon).

The two longest-serving directors are Sam K. Seymour, Jr., who served thirty-six years, from 1945 to 1981; and M. C. Dalchau, who served thirty years, from 1951 to 1981.

Two women have been appointed to the board: Patricia Wilson Scanlan in 1983 and Rita Myatt Radley in 1987.

Thomas C. Ferguson, one of the original board members, also takes credit for serving three different terms over the greatest length of time: he was appointed in 1935, 1945, and 1966.

The following chart lists the directors alphabetically and includes their home county, the official(s) who appointed them, their occupations, and their years of service on the board.

For many directors, the official expiration date of their term has been listed, although some directors may have remained in office pending the appointment of their successors. For the fifteen current directors, the official expiration dates of their terms are listed.

LCRA Board Membership
1935 to present

Director	County	Appointed by	Occupation	Years of Service
Anderson, Milton J.	Colorado	Clements	Farmer	2-12-81 to 1-1-87
Archer, William T., Jr.	Travis	Clements	Banker	8-30-79 to 2-22-84
Arnim, E. A.	Fayette	Stevenson Shivers Daniel Connally	Attorney	12-31-46 to 1-1-71
Arnold, William B.	Bexar	Atty. Gen. Allred	Publisher of labor paper	9-14-37 to 5-1-43
*Bandy, Lawrence Roy	Caldwell	Clements	Oil business	7-23-87 to 2-1-93
*Barker, Raymond F.	Kerr	Clements	Bank chairman	7-23-87 to 2-1-93
Brigham, Percy T.	Blanco	Shivers Daniel	Banker	2-2-53 to 1-1-65
Brooks, Raymond	Travis	Atty. Gen. McCraw	Newspaper reporter	1-1-35 to 1-1-47
Buttery, Orville	Llano	Stevenson	Druggist	5-3-43 to 1-1-49
Collins, Will	Llano	Land Com. Robinson	Newspaper reporter	8-10-35 to 1-1-37
Connally, John	Blanco	Land Com. Giles	Lawyer	3-18-41 to 1-1-47
Corder, W. D.	Burnet	Stevenson Shivers Daniel Connally	Merchant	10-24-45 to 8-7-73
Crider, Bower	Bastrop	Shivers	Lawyer	5-15-51 to 10-1-52

Crist, Charles E.	Blanco	Stevenson	Merchant	1-7-47 to 6-26-51
Dalchau, M. C.	Llano	Shivers Daniel Connally Briscoe	Merchant	5-15-51 to 1-1-81
Davis, T. H.	Travis	Allred	Banker	8-15-36 to 10-12-39
Dean, Tom	San Saba	Smith Briscoe	Rancher	1-4-73 to 1-1-85
*Dixon, John Kenneth	Llano	Clements	Specialty contractor	4-23-81 to 2-1-93
Ellis, Morris	San Saba	Clements	Rancher	8-3-82 to 1-1-83
Elzner, B. A.	Bastrop	O'Daniel	Merchant	5-5-41 to 1-1-47
Engelhard, Fritz	Colorado	Land Com. Walker	Farmer	1-1-35 to 3-1-41
Entzminger, Joe	Matagorda Connally	Connally	Lawyer	7-15-67 to 1-1-73
Ferguson, Thomas C.	Burnet	Atty. Gen. McCraw	Lawyer	1-1-35 to 1-1-37
		Stevenson		4-26-45 to 10-1-45
		Connally		2-3-66 to 1-1-71
		Smith		3-2-71 to 8-31-71
Fry, Roy	Burnet	Land Com. Walker	Druggist	1-1-35 to 1-1-45
*Garrison, James B., Jr.	Burnet	Clements	Bank chairman	7-23-87 to 2-1-93
*Grimes, James Randall	Williamson	Clements	Attorney	7-23-87 to 2-1-93
Hamill, P. R.	Matagorda	Stevenson	Banker	7-23-43 to 5-1-48
Hancock, John W.	Wharton	Connally Briscoe Clements	Banker, rice miller	5-13-63 to 1-1-87
Hardin, Alex	Llano	Jester	Rancher	12-30-48 to 1-1-55
Hodges, Morris	Colorado	Shivers	Lawyer	4-7-55 to 1-1-61
Huebner, Bert L.	Matagorda	Briscoe	Lawyer	1-15-79 to 1-1-85
Hutchins, J. F.	Wharton	Allred		1-17-36 to 5-1-41
Inks, Roy B.	Llano	Land Com. Walker	Merchant, businessman	1-1-35 to 8-5-35
Johnson, D. R.	Bastrop	Shivers	Merchant	4-7-55 to 1-1-61
*Johnson, Jack M.	Colorado	Clements	Agricultural businessman	7-23-87 to 2-1-93
Jones, John W.	McCullough	Clements	Banker	1-13-81 to 1-1-87
Jungmichel, Charles H.	Fayette	Smith	Insurance	1-4-73 to 1-1-79
Key, J. R.	Lampasas	Allred	Businessman	1-1-35 to 1-22-36
Kuykendall, Clay	San Saba	Stevenson Jester Shivers Daniel	Banker	5-4-43 to 1-1-67
*LeTulle, Burton B.	Matagorda	White	Real estate	3-4-85 to 12-31-90
Lewis, J. C.	Matagorda	Jester Shivers Daniel Connally	Banker, farmer / rancher, oil and gas	5-10-48 to 1-1-73
*Littlejohn, Jack	Fayette	White	Retail business	4-11-85 to 3-1-91

*Long, Cecil	Bastrop	Daniel Connally Smith White	Banker, merchant, farmer / rancher	2-3-61 to 1-1-79 1-31-84 to 3-1-91
Long, Robert	Travis	Smith	Lawyer	11-4-71 to 8-16-77
*Martin, James A.	Travis	White	Union representative	2-22-84 to 1-1-89
Matula, Charles	Fayette	Atty. Gen. McCraw	Farmer	1-12-39 to 5-1-45
*Matus, Charles	Blanco	Clements	Businessman	1-1-83 to 1-1-89
Mayer, E. B.	Fayette	Briscoe	Businessman	1-15-79 to 1-1-85
Mayfield, Eli	Matagorda	Smith	Lawyer	1-4-73 to 1-1-79
McLean, Martin	Burnet	Clements	Businessman	8-14-81 to 1-1-87
*Miller, Jack B.	San Saba	White	Attorney	1-1-85 to 1-1-91
Miller, Tom, Jr.,	Travis	Connally	Banker	2-3-66 to 1-1-71
Morgan, Guiton	Travis	Shivers	Lawyer	2-3-53 to 5-22-59
Morgan, J. Ed	Fayette	Daniel	Banker	5-22-59 to 1-1-65
Nash, John H., Jr.	Travis	Briscoe	Businessman	8-16-77 to 9-30-79
*Oles, Chas. Patrick, Jr.	Travis	Clements	Businessman	7-23-87 to 1-1-89
Parker, B. L.	Bastrop	Briscoe	Businessman	1-15-79 to 12-15-83
Parker, R. H., Sr.	Matagorda	Shivers	Businessman	5-15-51 to 1-1-53
Payne, John H.	Travis	O'Daniel	Newspaper publisher	10-31-40 to 5-1-43
Pennington, C. R.	Taylor	Allred	Businessman	1-1-35 to 6-1-45

Petri, Bill	Travis	Connally
		Smith
		Briscoe
			Linotype operator,
			  stereotyper, labor
			  representative
				2-3-66 to 8-30-79

*Radley, Rita Myatt	Wharton	Clements	Rancher, farmer, banker	7-23-87 to 2-1-93

Reinhard, A. J.	Tarrant	Atty. Gen.
		  McCraw
			Labor
			  representative
				1-1-35 to 1-1-39

Scanlan, John M.	Travis	Clements White	Lawyer	1-31-80 to 7-1-87
*Scanlan, Patricia W.	Fayette	Clements	Rancher	1-1-83 to 1-1-89
Schreiner, Charles III	Kerr	Briscoe	Banker, rancher	10-24-75 to 12-31-80
Selig, Marvin	Guadalupe	Clements	Businessman	4-23-81 to 12-31-86
Seymour, Sam K., Jr.	Colorado	Stevenson Shivers Daniel Connally Briscoe	Banker, merchant	5-17-45 to 1-1-81
Shapiro, Harry	San Saba	Connally Briscoe	Merchant	7-15-67 to 1-4-73 8-16-77 to 5-16-82
Schumann, Merritt	Comal	Clements	Insurance agent	1-31-81 to 1-1-87
Spires, A. B.	Travis	Stevenson	Banker	12-31-46 to 1-1-65
Strahan, Jake	Burnet	Briscoe	Merchant	11-15-73 to 1-1-81
Tate, Milton, Jr.	Washington	Briscoe	Lawyer	10-24-75 to 12-31-80
Thornton, J. R. "Bob"	Hays	Briscoe	Banker	10-24-75 to 12-31-80
Trigg, Jim Jones	Bastrop	Daniel	Lawyer	2-3-61 to 1-1-67
Voekel, Aubrey	Fayette	Smith	Banker	3-2-71 to 1-1-77
Waugh, Alex	Bastrop	Shivers	Businessman	2-3-53 to 4-6-55

White, Carl	Jefferson	Land Com. McDonald	Newspaper editor	1-4-37 to 7-23-43
Winters, Melvin C.	Blanco	Shivers	Lawyer	7-10-51 to 1-1-53
Wright, R. D.	Wharton	Stevenson Shivers Daniel	Banker	4-26-45 to 5-13-63
Yarborough, Ralph W.	Travis	Allred	Lawyer	2-9-35 to 1-7-36
Zercher, Roger Gilbert	Blanco	Smith Briscoe	Farmer	3-2-71 to 1-1-83

(*current members; terms to expire on dates shown)

Board of Directors
(In order of succession from 1935)

Brooks, Raymond	1-1-35 to 1-1-47	Lewis, J. C.	5-10-48 to 1-1-73
Spires, A. B. (Bryan)	12-31-47 to 1-1-65	Jungmichel, Charles	1-4-73 to 1-1-79
Miller, Tom, Jr.	2-3-66 to 1-1-71	Mayer, E. B.	1-15-79 to 1-1-85
Zercher, Roger G.	3-2-71 to 1-1-83	Littlejohn, Jack	4-11-85 to 3-1-91
Matus, Charles	1-11-83 to 1-1-89	Key, J. R.	1-1-35 to 1-22-36
Englehard, Fritz	1-1-35 to 3-1-41	Davis, T. H.	8-15-36 to 10-12-39
Connally, John	3-18-41 to 1-1-47	Payne, John H.	120-31-40 to 5-1-43
Crist, Charles E.	1-7-47 to 6-26-51	Buttery, Orville	5-3-43 to 1-1-49
Winters, Melvin C.	7-10-51 to 1-1-53	Hardin, Alex	12-30-48 to 1-1-55
Brigham, Percy T.	2-2-53 to 1-1-65	Hodges, Morris T.	4-7-55 to 1-1-61
Ferguson, Thomas C.	2-3-66 to 1-1-71	Trigg, Jim Jones	2-3-61 to 1-1-67
Long, Robert J.	11-4-71 to 8-16-77	Entzminger, Joe	7-15-67 to 1-1-73
Nash, John H., Jr.	8-16-77 to 9-30-79	Mayfield, Eli	1-4-73 to 1-1-79
Scanlan, John M.	1-31-80 to 7-1-87	Huebner, Bert L.	1-15-79 to 1-1-85
Oles, Charles P., Jr.	7-23-87 to 1-1-89	LeTulle, Burton B.	3-4-85 to 12-31-90
Ferguson, Thomas C.	1-1-35 to 1-1-37	Pennington, C. R.	1-1-35 to 6-1-45
Arnold, William B.	9-14-37 to 5-1-43	Wright, R. D.	4-26-45 to 1-1-63
Kuykendall, Clay	5-4-43 to 1-1-67	Hancock, John W.	5-13-63 to 1-1-87
Shapiro, Harry	7-15-67 to 1-1-73	Radley, Rita Myatt	7-23-87 to 2-1-93
Dean, Tom	1-4-73 to 1-1-85	Reinhard, A. J.	1-1-35 to 1-1-39
Miller, Jack	1-1-85 to 1-1-91	Matula, Charles	1-12-39 to 5-1-45
Fry, Roy	1-1-35 to 1-1-45	Seymour, Sam K.	5-17-45 to 1-1-81
Ferguson, Thomas C.	4-26-45 to 10-1-45	Anderson, Milton Jay	2-12-81 to 1-1-87
Corder, W. D.	10-24-45 to 8-7-73	Johnson, Jack	7-23-87 to 2-1-93
Strahan, Jake	11-15-73 to 1-1-81	Yarborough, R. W.	2-9-35 to 1-7-36
McLean, Martin E.	8-14-81 to 1-1-87	Hutchins, J. F.	1-17-36 to 5-1-41
Garrison, Jim	7-23-87 to 2-1-93	Elzner, B. A.	5-5-51 to 1-1-47
Inks, Roy B.	1-1-35 to 8-5-35	Arnim, E. A.	12-31-46 to 1-1-71
Collins, Will	8-10-35 to 1-1-37	Voelkel, Aubrey	3-2-71 to 1-1-77
White, Carl	1-4-37 to 7-23-43	Shapiro, Harry	8-16-77 to 5-16-82
Hamill, P. R.	7-23-43 to 5-1-48	Ellis, Morris	8-3-82 to 1-1-83
		Scanlan, Patricia Wilson	1-11-83 to 1-1-89

In 1951 an act was passed to increase the number of board members from nine to twelve:

Crider, Bower	5-15-51 to 10-1-52	Long, Cecil	2-3-61 to 1-1-79
Waugh, Alex	2-3-53 to 4-6-55	Parker, B. L.	1-15-79 to 12-15-83
Johnson, D. R.	4-7-55 to 1-1-61	Long, Cecil	1-31-84 to 3-1-91

Dalchau, M. C.	5-15-51 to 1-1-81
Dixon, Kenneth	4-23-81 to 2-1-93
Parker, R. H.	5-15-51 to 1-1-53
Morgan, Guiton	2-3-53 to 5-22-59

Board members increased to fifteen in 1975:

Tate, Milton Y., Jr.	10-24-75 to 12-31-80
Jones, John W.	1-13-81 to 1-1-87
Barker, Raymond F.	7-23-87 to 2-1-93
Thorton, J. R. "Bob"	10-24-75 to 12-31-80
Selig, Marvin	4-23-81 to 1-1-87

Morgan, J. Ed	5-22-59 to 1-1-65
Petri, Bill	2-3-66 to 8-1-79
Archer, William T., Jr.	8-30-79 to 2-22-84
Martin, James A.	2-23-84 to 1-1-89

Bandy, Lawrence Roy	7-23-87 to 2-1-93
Schreiner, Charles III	10-24-75 to 12-31-80
Schmann, Merritt	1-13-81 to 1-1-87
Grimes, Randy	7-23-87 to 2-1-93

Bibliography

Books

Austin American-Statesman Staff. *Fellow Texans in Profile.* Austin: 1948.

Banks, Jimmy. *Money, Marbles and Chalk — The Wondrous World of Texas Politics.* Austin: Texas Publishing Co., Inc., 1971.

Brown, John Henry. *History of Texas.* 2 vols. St. Louis: L. E. Daniel Publishers, 1893.

Burke's Texas Almanac. Houston: 1881.

Caro, Robert A. *The Path to Power.* New York: Alfred A. Knopf, Inc., 1982.

Chesser, Allen H. *Let There Be Light — History of Guadalupe Valley Electric Cooperative, 25th Anniversary.* San Antonio: Taylor Company, 1964.

Childs, Marquis. *The Farmer Takes A Hand.* Garden City, NY: Doubleday & Company, 1952.

Clay, Comer. *The Lower Colorado River Authority, Public Administration and Policy Formation.* Austin: University of Texas Press, 1956.

Davis, William J. (ed.) *The Partisan Rangers — Memories of Gen. Adam R. Johnson.* Louisville, KY: G.G. Fetter Company, 1904.

Ellis, Clyde T. *A Giant Step.* New York: Vantage Books, 1966.

Fehrenbach, T. R. *Lone Star, A History of Texas and Texans.* New York: Macmillan Publishing Company, 1968.

Frizell, Joseph F. *Report on the Proposed Dam and Water Works for the City of Austin, Texas, and Report of Consulting Engineer.* Austin: Von Boeckmann-Jones Company, 1892.

Henderson, Richard B. *Maury Maverick, A Political Biography.* Austin: University of Texas Press, 1970.

Hendrickson, Kenneth E., Jr. *The Waters of the Brazos, A History of the Brazos River Authority, 1929–1979.* Waco: Texian Press, 1981.

Humphrey, David C. *Austin, An Illustrated History.* Austin: Windsor Publications, Inc., 1985.

Johnson, Sam Houston. *My Brother Lyndon.* New York: Cowles Book Company, Inc., 1969.

Lilienthal, David E. *TVA, Democracy on the March.* New York: Harper & Brothers, 1944.

Long, Walter E. *Flood to Faucet.* Austin: The Steck Company, 1956.

Mooney, Booth. *The Lyndon Johnson Story.* New York: Farrar, Strauss & Company, 1956.

Reed, S. G. *A History of Texas Railroads*. Houston: St. Clair Publishing Company, 1941.

Steighorst, Junann A. *Bay City and Matagorda County — A History*. Austin: Pemberton Press, 1965.

Steinberg, Alfred. *Sam Rayburn*. New York: Hawthorn Books, 1975.

Texas Almanac and State Industrial Guide. Various editions. Dallas: A.H. Belo Corporation.

Newspapers and Periodicals

Abilene Reporter-News
Austin American
Austin American-Statesman
Austin Daily Statesman
Austin Daily Tribune
Austin Statesman
Austin Weekly Statesman (October 31, 1899)
Bay City Tribune (Century of Progress edition, August 26, 1945)
Burnet Bulletin
Chicago Daily Tribune
Dallas Morning News
Dallas Times-Herald
Eagle Lake Headlight (August 12, 1938)
El Paso Herald-Post
Fort Worth Star-Telegram
Highland Lakes News
Houston Chronicle
Houston Post
Johnson City Record-Courier
LCRA Currents
LCRA News
LCRA Newsletter
LCRA River Review
Lubbock Avalanche-Journal
Marble Falls Highlander
New York Herald-Tribune
San Angelo Evening Standard
San Angelo Standard-Times
San Antonio Express
State Observer
Texas Co-op Power
Texas Observer
Texas State Gazette (October 9, 1852)
Tri-weekly State Gazette (July 14, 1869)
U.S. News & World Report (February 20, 1967)
Waco News-Tribune
West Texas Today (West Texas Chamber of Commerce)

Articles, Reports and Unpublished Manuscripts

"Annual Report of the Secretary of War, 33rd Congress, Second Session." Executive Document II (1854).

Bolton, Herbert E. "Location of LaSalle's Colony on the Gulf of Mexico." *Southwestern Historical Quarterly* (April 1949).

Brown, Frank. "Annals of Travis County and the City of Austin, from the Earliest Times to Close of 1875." (Manuscript and typed copy in Austin-Travis County History Collection, Austin Public Library.)

Bunger, Howard P. "Irrigation and Flood Protection, Austin to Matagorda, Texas." Report No. 20, U.S. Bureau of Reclamation (1936).

"By the People, For the People — The Story of Rural Electric Corporations of Texas." Department of Agricultural Education, Texas A&M College, 1966.

"Camp Swift, Texas," by Oscar Park Houston and Walter E. Long. (Unpublished manuscript in Austin-Travis County History Collection.)

Clay, Comer. "The Lower Colorado River Authority — A Study in Politics and Public Administration." Ph.D. dissertation, University of Texas, June 1948.

"Development of Texas Rivers, A Water Plan for Texas." Austin: Texas Planning Board, 1938.

Dowell, C. L. "The Lower Colorado River Authority." *Public Power Magazine* (February 1945).

Edwards, Pauline. "The Lower Colorado River Authority, An Agency of the State." Circa 1982.

Fanning, J. T. "Report on the Public Water Supply and Electric Lighting of the City of Austin." Austin: Von Boeckmann-Jones Co., 1892.

"Flood Control on Colorado River, Texas." 66th Congress, 1st Session, House Document 304 (1919).

Hawley, Freese and Nichols, Fort Worth. "Power Market Study for LCRA, Austin, Texas." March 1938.

Haynie, R. W. "President's Report to the 10th Annual Convention of the West Texas Chamber of Commerce." *West Texas Today* (June 1929).

"Highland Lakes of Texas, The." National Parks Service, U.S. Department of Interior, 1941.

Hill, Guy Cummings. "The History and Purpose of the Lower Colorado River Authority." Master's thesis, University of Texas, June 1935.

"Improvement of the Colorado River from Austin to the Gulf." Folder published by the Colorado River Improvement Association. Austin: 1915.

Johnson, William H. "A Short History of the Sugar Industry in Texas." Texas Gulf Coast Historical Association, *Publication V* (April 1961).

"Lower Colorado River Reclamation Study." State Reclamation Department. Austin: 1936.

Lowery, B. L. "Flood Control by Marshall Ford Reservoir." Washington: 1937.

Mahon, Ralph J. "Report of Colorado River Flood of July-August, 1938." State Reclamation Department. Austin: September 19, 1938. (Report to Texas Senate Interim Investigating Committee.)

"Marshall Ford Dam, Colorado River Project, Texas — Report on Allocation of Construction Costs." Washington: U.S. Department of Interior (August 1947).

"Marshall Ford Development, Justification for High Dam, Colorado River Project in Texas." E. B. Debler and John H. Ritter, 1937.

McDonald, John. "The Great Dam and Water and Light System at Austin, Texas." Austin: 1893.

McLean, Roy J. "Proceedings of Convention on Colorado River Flood, Saturday Morning, July 30, 1938."

Mead, Daniel W. "Report on the Dam and Water Power Development at Austin, Texas." Madison, WI: 1917.

Randolph, Beverley. "History of Predecessor Companies." Austin: 1974.

Schmidt, Lewis A. "A Two-Hundred Foot Hydro-Electric Development at Marshall Ford Site on Colorado River." Master's thesis, University of Texas, June 1933.

Stranahan, Harris & Co., Inc. "Bond Prospectus — $21,635,000 Lower Colorado River Authority, Texas, Revenue Bonds." New York: May 27, 1943.

"Surface Water Supply and the United States — 1945." Water Supply Paper 1038. Washington: 1947.

Taylor, T. U. "The Austin Dam." Water Supply Paper 40. Washington: 1900.

———. "Irrigation Systems of Texas." Water Supply Paper 71. Washington: 1902.

———. "The Water Powers of Texas." Water Supply Paper 105. Washington: 1904.

"Texas Floods of 1938 and 1939." S. D. Breeding and Tate Dalrymple. Water Supply Paper 914. Washington: 1944.

Weiner, William. "Rate-Making Policy and Financial Obligations of the Lower Colorado River Authority." Master's thesis, University of Texas, August 1946.

Williams, B. F. "Report of Inspection of the Lower Colorado River from Wharton to Bay City." May 24, 1928.

Wintermann, David R. "History of Rice in Eagle Lake Region." Circa 1957. (Typewritten manuscript in the Archives Section, Eula and David Wintermann Library, Eagle Lake, Texas.)

Miscellaneous

Congressional Record.
General and Special Laws of Texas.
House Journals (Texas).
LCRA Board of Directors, Minutes.
Revised Civil Statutes of Texas.
Senate Journals (Texas).
Texas Board of Water Engineers, Minutes.
U.S. Statutes at Large.

This drawing by Adam Johnson in the 1850s is the earliest known proposal for a dam on the Colorado River. Johnson proposed building a dam at the site where Buchanan Dam is located.

— Photo courtesy Dr. Comer Clay

Construction workers on the original Austin Dam, built 1890–93.

The Austin Dam formed Lake McDonald, a source of hydroelectric power and recreation for Austin. Pictured is the Ben Hur, *a large paddlewheel steamboat.*

The flood of 1900 destroyed the Austin Dam. Floodwaters crested forty feet over the dam, eventually pushing out the middle section.

Also destroyed were the Ben Hur *and the city's hydroelectric plant.*

The City rebuilt the Austin Dam, amid the rubble of the original structure.

The second Austin Dam, completed in 1909, was almost immediately damaged by flooding.

The narrow floodgates atop the second Austin Dam may have actually contributed to the dam's destruction. Debris built up against the gates and blocked the discharge of floodwaters.

Twice damaged by floods, the second Austin Dam remained in this condition for more than twenty years.

A soldier stands guard over what is left of the water in Lake Austin after releases have been made for downstream irrigation needs. This picture was taken in 1918.

— Photo courtesy Austin History Center

The front page of the Austin American, *Saturday, November 10, 1934, anticipates the passage of the bill creating LCRA (sometimes referred to as the Colorado Valley Authority or the Colorado River Project). The Texas legislature had defeated the bill three times before approving it on its fourth try.*

Construction on Hamilton (later Buchanan) Dam, at that time controlled by the Insull utility interests. This photo was taken in 1932, shortly before Insull declared bankruptcy. Hamilton Dam was about forty percent complete.

A houseboat washes over the Austin Dam in 1935.

A June 1935 aerial shot of downtown Austin, where Congress Avenue is inundated by flooding.

Downtown Austin during the 1936 flood, looking north toward the Capitol from Congress Avenue.

Dedication ceremonies for the third Austin Dam, April 6, 1940 — one day shy of the fortieth anniversary of the destruction of the original structure. LCRA rebuilt the dam, which was renamed the Tom Miller Dam in honor of the mayor of Austin.

An aerial shot, taken in the late 1930s, shows construction of Mansfield Dam.

LCRA sponsored "electric fairs" during the late 1930s and early 1940s to acquaint Central Texas families with the advantages of electricity in the home.

Lake Travis during a dry spell in July 1978. Lack of rainfall increases the demand for water, which lowers the elevations of Lakes Buchanan and Travis.

Starcke Dam discharges floodwaters into Lake Travis during a flood in September 1952. The level of Lake Travis rose by more than fifty feet within a twenty-four-hour period.

Members of LCRA's first board of directors pose with LCRA supporters in front of the Capitol in 1935. Included are Roy Inks (third from right), for whom Inks Dam and Lake are named; Thomas C. Ferguson (far left), for whom the Ferguson Power Plant is named; and Ralph Yarborough (sixth from right), who later served as U.S. senator from 1957 to 1971.

Beverley Randolph (left) prepares for a flight to Llano County in 1931 to secure land rights for Hamilton (later Buchanan) Dam. She was secretary to Alvin Wirtz and later served as head of LCRA's Office Services department.

Portrait of Governor Miriam A. "Ma" Ferguson, whose support of the LCRA bill assured its passage by the Texas legislature.

U.S. Representative J. P. "Buck" Buchanan, who lobbied President Roosevelt for federal funds to complete the Hamilton Dam project. LCRA renamed the dam in his honor.

Ralph Yarborough, photographed around 1935.

Alvin Wirtz, who wrote the bill creating LCRA, served as the authority's first general counsel.

President Roosevelt is pictured with newly elected U.S. Representative Lyndon Johnson and Governor James Allred at Galveston in April 1937. Roosevelt and Johnson formed a friendly working relationship that would prove beneficial to LCRA.

— Photo courtesy LBJ Library

Lyndon Johnson, photographed at a 1939 meeting of the newly formed Pedernales Electric Cooperative. The handwritten inscription by Johnson is to LCRA General Manager Max Starcke.

U.S. Representative Lyndon Johnson kneels beside U.S. Representative J. J. Mansfield, for whom Mansfield Dam was named, during dedication ceremonies in 1941. Standing at far left is Alvin Wirtz; fourth from right is General Counsel Sim Gideon; at far right is LCRA General Manager Max Starcke.

LCRA Board of Directors, at a meeting in the early 1940s. John Connally, director from Blanco County, is sitting in the corner; General Counsel Sim Gideon is second from right; and General Manager Max Starcke is at right.

LCRA management, late 1950s, poses for this photo. Left to right are John Kight, treasurer; R. A. Lucksinger, controller and assistant general manager; Sim Gideon, general manager; G. E. Schmitt, chief engineer and assistant general manager; and Mac Umstattd, general counsel.

Sim Gideon at an LCRA appreciation dinner, hosted by the City of Austin, for LCRA's role in controlling the 1957 flood. Governor Price Daniel and Lady Bird Johnson are seated at Gideon's left.

Clarence McDonough (1935–40)

Max Starcke (1940–55)

Sim Gideon (1955–73)

Charles Herring (1973–81)

Elof Soderberg (1981–86)

S. David Freeman (1986–present)

Buchanan Dam / Lake Buchanan

Inks Dam / Lake Inks

Wirtz Dam / Lake LBJ

Starcke Dam / Lake Marble Falls

Mansfield Dam / Lake Travis

Tom Miller Dam / Lake Austin

Sim Gideon Steam Plant, Bastrop

T. C. Ferguson Power Plant, Marble Falls

Irrigation canals, Matagorda County. LCRA's irrigation divisions water as much as 72,000 acres in Colorado, Wharton, and Matagorda counties. These counties produce up to twelve percent of the state's rice crop.

Fayette Power Project, La Grange

Index

A

Alabama Power Company vs. Ickes, 92
Alamo, 63
Alamo National Bank, 72
Alexander, Charles H., Jr., 28, 31, 75
    Charles H., Sr., 28
    Jay, 28, 31, 32
Allred, Dave, 205
    James, 51–52, 63, 64, 71, 96, 100–
    101, 114–115
Alsop, R. B. (Bob), 81, 93, 94
Alvin J. Wirtz Dam, 113, 179–180, 231
    (*see also* Granite Shoals Dam)
Anderson, Art, 143–146, 153, 156
    Jay, 221, 227
Annals of Travis County and the City of Austin,
    11, 18
Arnim, E. A. (Sam), 114–115, 191
Arnold, Ward S., 31, 32
    William B., 160
Associated Engineers, 32
Atlantic Richfield, 209
Atwood, Roy, 165
Austin, Moses, 5–6
    Stephen F., 5–6, 10
Austin Board of Trade, 19
Austin Building Company, 207
Austin Chamber of Commerce, 24, 29,
    45, 95, 151, 159
Austin City Council, 25, 26
Austin Dam, 19–26, 82, 94, 95, 103,
    113–114, 150, 151 (*see also* Tom
    Miller Dam)
Austin Dam, Inc., 26
Austin Housing Authority Board, 189

Austin—An Illustrated History, 22
Austin North-Western Railroad, 25
Austin Rapid Transit Company, 23
Austin, Texas, 8, 9, 11, 19–26, 57, 70,
    121, 166, 171, 178–179
Austin Water, Light and Power
    Company, 20, 21
Autobiography of a Curmudgeon, 80
Avery, C. N., 100, 101

B

Babcock, John, 177, 178
Bagalay, John, 220, 222
Baker, L. Buck, 165
Ballinger, Texas, 219
Bartlett, Texas, 129
Bashore, Harry W., 121, 147
Bastrop, Texas, 11, 154, 159
Bay City Irrigation Company, 14
Bay City, Texas, 15, 186
Bechtel Company, 207
Bell County, 129
Bellville, Texas, 154
Ben Hur, 22–23, 25
Bentley, Max, 52
Bernhard, Harry, 165, 166
Betty Powell, 9
Bishop, Joseph E., Jr., 84
Blanco, Texas, 139
Bluebonnet Electric Cooperatives, 200–
    201, 202, 225
Board of Public Works (Austin), 20–21
Bogart, John, 20
Bonneville Power Administration, 128
Boughton, I. C., 36
Boulder Dam, 46

Bowers vs. City of Taylor, 164
Boynton, Alexander, 122
Brazos River, 5, 6
Brazos River Authority, 110–111
Brazos River Authority bill, 62
Brazos River Conservation and
    Reclamation District, 88
Brazos River District, 51
Brees, Hubert J., 159
Brenham, Texas, 154
Brooks, S. Raymond, 65, 66, 67, 71, 161
Brown, Frank, 11, 18
    George, 31, 82, 97, 98
    Herman, 31, 82, 98
    John Henry, 4, 5
Brown & Root, Inc., 82, 95, 97–98, 102,
    148, 206, 207–210
Brown County Water Control and
    Improvement District, 14
Brown County Water Improvement
    District No. 1, 28
Brownlee, Houghton, 96, 100, 101, 103
Brownwood Chamber of Commerce, 50
Brownwood Dam, 14–15, 29
Buchanan, James Paul, 30, 38, 41, 45,
    46, 58, 59–60, 65, 68, 71, 76–77,
    83, 92, 97, 98, 102
    James Paul, Jr., 98
    Mrs. James Paul, 103
Buchanan Dam, 27, 38, 45, 54, 55, 65,
    71, 87, 89–90, 92–93, 94, 114,
    120–121, 123, 138, 231 (*see also*
    Hamilton Dam)
Burke, ———, 58
Burnet Chamber of Commerce, 50
Burnet County, 38, 45, 129
Burnet, Texas, 39, 139, 154
Buzzard Falls, 89

C

Cain, Stephanie, 180
Camp Swift, 159–161
canal companies (*see* irrigation
    companies)
Cárdenas y Magaña, Manuel José de, 5
Carlton, Leonard, 166
Carmody, John, 132
Carney, Stephen J., 177
Caro, Robert A., 97, 107, 131
Carpenter, John W., 139–141
Carrington, John B., 29

Carter, Jimmy, 229
Caswell Cotton Mill, 25
centennial bill, 50, 59, 63
Central Power and Light Company, 29,
    33, 34, 37, 126, 139, 142, 152
Central Texas Electric Cooperative,
    Inc., 142, 187
Central Texas Hydro-Electric
    Company, 33, 34–36, 45, 46, 75
Chadwick, Milam, 19
Chamberlain, F. G., 43
    Tom, 28
Chautauqua, 22
Childress, Texas, 88
Church and Graves, 44
Civil War, 9
Clark, Edward A., 101, 199
    Lee C., 70, 72, 99
Clay, Comer, 11, 15, 25
Coastal States Gas Producing
    Corporation, 196, 200, 202, 203–
    204, 205, 225
Cofer, Hume, 203
Collie, Wilbourne B., 120
Collins, Will, 74, 85–86
Colorado, 9
Colorado County, 14
Colorado River: agricultural uses of, 3,
    4, 10; clearing begins, 7; described,
    1–3; early exploration along, 4–5;
    first appearance on maps, 5; first
    attempts to dam, 19–26, 27–28,
    29–37; first government aid to
    improve, 7–8; flooding of, 3, 6–7,
    8–9, 11 (worst), 16, 24, 25–26, 46,
    70, 90, 91, 114, 119–125, 174, 186;
    navigation of, 3, 6–7, 8–9, 91;
    normal flow of, 4; other names for,
    5; "raft" of, 5, 6–7, 8–9, 15–16;
    recreation along, 24–25; settlers
    along, 5–7, 10; surveys of, 28, 30;
    water rights to, 12, 14, 17, 28–29,
    30, 47, 48, 52–54, 58, 62, 103
Colorado River Authority: bill to
    establish, 40, 45–50, 51–61;
    constitutionality of, 65–66, 68; *see
    also* LCRA
Colorado River Company, 36, 37, 45, 46,
    47–48, 56, 58, 67, 68, 70, 74–75
Colorado River Improvement
    Association, 24, 30, 41, 45, 119

Colorado River Municipal Water
  District (CRMWD), 219–223
Colorado River Navigation Company, 7
Colorado River Valley Area Committee,
  55
Colorado Valley Authority, 59
Comal Power Plant, 187
Community Public Service Company,
  88
Concho River, 3
congressional districts, 38
Connally, John B., 168, 188–191, 197,
  199, 234
  Tom, 38, 41, 45, 115, 116
Cook, Morris, 129
Corcoran, Thomas G. (Tommy the
  Cork), 132
Corder, W. D., 198
Corrigan, Bernard, 20
Coughlin, J. R., 84, 86
Craddick, Tom, 220
Craig, Richard, 198
Cranford, C. M., 165
Crockett, Moton, 183
Crowley, Louisiana, 12
Crump, M., 72–73
Cuero, Texas, 142
Cureton, C. M., 64

D

Dalchau, M. C., 198, 234
Dale, F. A., 33
Dallas Federal Reserve Bank, 72
Dallas Morning News, 175
Dallas, Texas, 50, 63
Dancer, Jonas, 18
Daniel, Price, 187, 195
Davant, Will, 16
Davis, Billy F., 177
  T. H., 96
  W. C., 65, 66
Dean, Tom, 198, 212, 215
  W. V., 47, 48, 54
Debler, E. B., 124
Decker Coal Company, 224
Dibrell, J. B., 136
Dickard, Reagan, 32
Dixon, Ken, 227
Dorbandt, Christian, 49
drouths, 185–187
Duggan, Arthur, 52

*Duke Power Company vs. Greenwood County,
  South Carolina*, 92
Duke Power and Light Company, 88
Dunovant, William, 12

E

E. V. Spence Reservoir, 221
Eagle Lake, Texas, 12
Edison, Thomas Alva, 29
Edison General Electric Company, 29
Edison Hotel, 145
Edna, Texas, 142
Edwards Plateau, 3
Eidelbach, Earl, 12–15
Eilers, A. J., 30
El Camino Real, 6
El Campo, Texas, 142
Electric Bond and Share Corporation
  (EBASCO), 88
Elephant Butte, 46
Elgin, Texas, 139
Elkins, Bess, 46
Emergency Relief Appropriation Act,
  83, 96
Emery, Peck and Rockwood
  Development Company, 31–32,
  33, 81
Engelhard, Fritz, 46, 52, 58, 60, 65, 67,
  99
Environmental Protection Agency, 211
explosives, 13
Exxon Coal Resources, USA, 217

F

Fanning, J. T., 20
Fargo Engineering Company, 32, 35, 72,
  75
Farmers State Bank of Seguin, 137
Fayette Power Plant, 207–210
Fayette Power Project, 204, 217
Federal Bureau of Investigation, 83–86
Fegles Construction Company, Ltd., 33,
  34, 35, 75, 83
Fehrenbach, T. R., 5, 6
Fellow Texans in Profile, 137
Ferguson, James E. (Pa), 3, 9, 48–49,
  54–55, 59, 63
  Miriam A. (Ma), 3, 9, 41, 48–49, 51,
  222
  Thomas C., 39–40, 49–50, 64, 65,
  66, 68–70, 72, 76–78, 81, 89, 91,

99, 102, 132, 149, 191, 205–206, 222, 234
Ferguson Power Plant, 196, 204–206
Flood to Faucet, 8, 24
Folsom, "Big Jim," 176
Foreman, Wilson, 196
Fortas, Abe, 148
Fort St. Louis, 5
Foster, Troy, 187
Foundation Company of New York, 80, 81
Fowler, Wick, 175
Frederick, John H., 159
Fredericksburg, Texas, 139, 141–142, 186
Freeman, S. David, 229
Frizell, Joseph P., 20
Fry, Roy, 65, 67, 71, 75, 78, 79, 89–90, 96, 129
fuel crisis, 184
Fuerst, E. S., 84
Fulcher, Gordon, 41–42

G

Gainesville, Texas, 88
Galloway, W. C., 39
Garwood Irrigation Company, 31, 103
General Land Office, 4
Georgetown, Texas, 138
German, S. H., 70
Giddings, Texas, 154
Gideon, Sim, 34, 43–44, 48–49, 62, 72, 107, 137, 150, 162–163, 170–171, 174, 184–185, 187–188, 191–193, 195, 200, 205, 222
Gideon Steam Plant, 187, 195–196, 201
Glenn, J. W., 19
Gonzales, Texas, 139, 142
Grand Coulee, Washington, 95
Granite Shoals Dam, 179–180 (*see also* Alvin J. Wirtz Dam)
Graves, Harry N., 52, 58
  Justice, 103
Green, Charles E., 161, 188
  Glenn M., Jr., 172
Grote, Henry, 130
Guadalupe-Blanco River Authority, 162, 172
Guadalupe River, 31
Gulf Coast Irrigation System, 14
Gulf Coast Water Company, 14, 102–

103

H

H. B. Zachry Company, 206, 207, 217
Hamill, P. R., 165, 168
Hamilton, G. W., 32
Hamilton Dam, 32–38, 44–45, 48, 66–67, 71, 74 (*see also* Buchanan Dam)
Hamlin, James D., 53
Hammond, Bob, 146
Hancock, John, 197, 198, 203, 207, 209, 215, 217
  R. L., 216, 225, 227
Harley, G. F., 150
Harris, Merton, 100, 101
Hatcher, Mary, 178
Hawley, Freese and Nichols, 109
Heaton, John D., 216
Hedges, Daniel, 228
Hendrickson, Kenneth E., Jr., 110
Herring, Bob, 202
  Charles F., 188–189, 192, 193, 196–200, 201–203, 206–207, 209, 210, 212, 228
Highland Lakes, 4, 17, 175–179, 194
Highland Lakes News, 175, 176
Hill, F. W., 160
  Joe, 120
Hill Country, 3, 108, 111, 127, 129, 141, 172
History of Texas, 1685–1892, 4, 5
History of Texas Railroad, A, 10
Hobby, Bill, 221
Hohn, Caesar "Dutch," 172
Holbrook, T. J., 52, 118, 120
Hollamon, Tom, 41–43
Hopkins, Harry, 102
  Welly K., 52
Hornsby, John, 52
Houston, Oscar Parke, 161
Houston Natural Gas, 202
Hughes, Sarah T., 38, 47–48, 50, 54, 63
Humble Oil and Refining Company, 31
Humphrey, David C., 22
Hunt, F. S., 41–43
  Henry T., 46, 53, 57, 58, 69, 76, 77–79
  Ralph W., 83–86
Huntsville, Texas, 88
Hurley, E. N., 36–37
Hutchins, J. F., 86, 103
hydroelectric power, 67–68, 88–92

I

IBEW-520, 165–166
Ickes, Harold, 37, 67, 69, 71, 77–80, 84, 95–96, 103–105, 148
illegal aliens, 206
Inks, Jim Moss, 103
  Mildred, 103
  Mrs. Roy, 103
  Roy B., 65, 71, 74, 81, 87
Inks Dam, 32, 92, 94, 138, 231
Inks Lake, 231
Insull, Martin J., 29–31, 33–35
  Samuel, 29–30, 33–35, 52
International Brotherhood of Electrical Workers, 113, 163–169
International and Great Northern Railroad, 25
Interstate Public Service, 14
irrigation companies, 12–17, 31, 62, 102–103
Ivie, Owen, 220, 222

J

Jerome Park Reservoir, 178
Jester, Beauford H., 169
Jim Wells County, 173
Johnson, Adam Rankin, 27–28, 32, 40, 81
  Lady Bird, 222
  Lyndon Baines, 1, 31, 38, 44, 68–69, 78, 82, 99–104, 107–109, 115–117, 122, 129–134, 139, 141, 147, 151, 152, 171–173, 180, 181, 187, 195, 199, 201, 229–230
  R. E., 40
  Sam Ealy, Jr., 107
  William D., 25, 26
Johnson City, Texas, 40, 141–142, 225–226
Jones, Grant, 220
  Herman, 205
  John W., 220
Jungmichel, Charles, 198

K

Karankawas, 4, 5
Kate Ward, 8–9
Kennedy, John F., 38
Kerrville, Texas, 154
Key, J. R., 64, 71

Kingsland, Texas, 17
Kleberg, Richard, 38, 69
Krick organization, 186
Kurtz, L. A., 161
Kyle, Texas, 139

L

labor strike, 165–169
La Grange, Texas, 11
Lake Austin, 233 (*see also* Lake McDonald)
Lake Brownwood, 14
Lake Buchanan, 4, 17, 114, 231
Lake Lyndon B. Johnson, 17, 179, 231
Lake McDonald, 21, 22, 24–25
Lake Marble Falls, 232
Lakeside Irrigation Co., Inc., 103, 217
Lake Travis, 4, 17, 185–186, 232
Lamar, Mirabeau B., 18
Lampasas, Texas, 139, 154, 168
Laredo Power and Light Company, 47
Lareno, 9
la Salle, Réne Robert Cavelier, Sieur de, 4
LCRA (Lower Colorado River Authority): acquires rural lines, 152; allowed to build steam plants, 162–163, 174, 184–185, 187, 195–196; begins operations, 2; buys Hamilton Dam, 74–75; coal-fired systems of, 204, 209–210, 224; created, 61–62; environmental record of, 211; fiftieth birthday of, 221–222; first operations manager, 133–135; first employee, 65; first general manager, 80; flood control measures of, 81–82, 83, 89–90, 91, 96, 102, 114–117, 119–125, 146–149, 174, 185, 187, 210–211; funding of, 4; issues revenue bonds, 62, 67, 69–70, 82; gas-fired units of, 196, 200, 204–205; general managers of, 3, 34, 80, 151, 157, 185, 193, 196–198, 210, 212–216, 228–229; income of, 4; irrigation services of, 53, 55–56, 89, 96, 103; and labor unions, 113, 157–158, 163–169, 207; land acquisition for, 95; lignite reserves of, 204, 210, 217, 219, 225, 227, 229; as power supplier, 160; publicity of, 174–179; recreational

aspects of, 155, 179, 190, 193, 214; responsibilities of, 62; and sale of electricity, 88–92, 108–112, 119–120, 121–122, 124, 126, 132–133, 139–142, 148, 151–157, 163, 187–188, 202–204, 226; and soil conservation, 171–174; in suit with Decker Coal, 224–225; in suit with utility companies, 88–92; use of alternate fuel sources, 204–205, 210, 218; wastewater plants of, 218, 219; water quality program of, 217, 218; *see also* LCRA bill, LCRA Board of Directors, Colorado River Authority
LCRA Act, 163, 171
LCRA bill: water rights amendment to, 47, 48, 52–54, 58, 60; *see also* Colorado River Authority
LCRA Board of Directors: appointments to first, 3–4; charged with negligence, 114–118, 119–125; first meeting of, 67; membership, *see* Appendix B; membership increase, 198; original members of, 64–65; review of, *see* Appendix B
León, Alonso de, 5
LeTulle, Burton, 227
   Victor L., 13–14, 15–16
Lewis, Gib, 220, 221
Liberty, Texas, 88
Lindley, Walter C., 36–37
Lipans, 4
Littlejohn, Jack, 227
Llano River, 3, 114
Llanos, Francisco de, 5
Llano, Texas, 139, 154, 186
Lometa, Texas, 19, 154
Lone Star, A History of Texas and Texans, 5, 6
Long, Cecil, 198, 227
   Robert, 198
   Walter E., 8, 24, 159
Looney, Everett, 199
Looney and Clark, 199
LoVaca Gathering Company, 200, 202, 203–204
Lower Colorado River Electric Cooperative, Inc. (LCREC), 151–152
Lower Colorado Valley, 3

Lucksinger, R. A., 93–94, 185, 191–192, 193, 203, 205
Luling, Texas, 142
Lynch, William, 139

M

McAllister, Walter, 28
McCann, Bill, 218, 228
McClellan, Carole, 210
McCrary, Tex, 178
McCraw, William, 65, 69, 206
McCulloch, C. A., 36–37
McDonald, John, 19, 21, 22
McDonough, Clarence, 80–83, 84–85, 89, 91, 93–94, 96, 113, 124, 133, 143, 144, 146, 150–151, 174
McGregor, T. H., 52, 58
McKee, Harry, 46
McKenzie Construction Company, 95
McLean, Martin, 227
   Roy J., 115
McMahon, R. J., 123
McMillan, Robert J., 35, 36
McWilliams, W. C., 187
Magnolia Petroleum Company, 31
Malott, C. G., 36, 67, 68, 69, 70, 71, 75–79
Mangum, Joe, 16
Manley, Chesley, 47
Mann, Gerald, 164
Mansfield, Joseph J., 30, 38, 45, 46, 101, 102, 115, 116, 149, 155
Mansfield Dam, 94, 155–156, 186, 232 (*see also* Marshall Ford Dam)
Marble Falls Dam, 179–180, 194–195 (*see also* Starcke Dam)
Marble Falls, Texas, 27, 40, 183
Markham, T. M., 70, 83
Marsh, Charles E., 79, 100–101
Marshall Ford Dam, 87, 94–98, 101–102, 116, 120, 123, 124, 138, 147, 150, 155–156 (*see also* Mansfield Dam)
Martin, Jack, 227
   Will R., 52
Matagorda Bay, 4, 16
Matagorda County, 12, 14, 15
Matagorda Peninsula, 16
Mathews, Charles, 225
Matthew Bayou, 12
Matus, Charles, 227

Matustik, David, 228
Maverick, Maury, 104
Maxey, C. J., 84
Mayer, E. B. (Tex), 212
Mayfield, Eli, 198
Mead, Daniel W., 26
Mercury Construction, 207
Metcalfe, Penrose, 54
Middle West Utilities Company, 29, 36
Midkiff, Morris, 121
Milam Building, 44
Miller, Jack, 227
   Tom, 82, 103, 121, 168, 180
   Tom, Jr., 191
Miller Flying Service, 32
Mississippi Valley Utilities Company, 33, 36, 37
Mississippi Valley Utilities Investment Co., 56
Moccasin Belle, 9
model of Highland Lakes area, 143–146
Montford, John, 220
Moody, Dan, 44, 124
Moore, Weaver, 54
Morrison, Ralph W., 36, 37, 46–48, 54, 58, 75–76
Morrison-Knudsen, Inc., 92
Moulton, Texas, 139, 142
Moursund, A. W., 201
Mundt, Leroy, 144–146

N

Nalle, Joseph, 19
Napier, ———, 72–73
National Industrial Recovery Act (NRA), 37, 89
National Labor Relations Act, 164
National War Labor Relations Board (NWLRB), 165
natural gas supplies, 200–203
Nieswanger, E. B., 139
Nixon, Richard, 229
Nordell, Mrs. Edward F., 177
Norris, John A., 65, 110
   Madalene, 65

O

O'Daniel, W. Lee (Pappy), 122, 199
O'Dwyer, William, 176
Ogle, Charles, 177
Old Three Hundred, 6
Operation Waterlift, 175–179
Orange Grove, Texas, 137

P

P.K. Williams Nash Company, 177
Pace Bend Park, 193, 214
Parker, O. D., 24
Parkhouse, George, 50, 63
Parr, Archer, 52
   George, 199
Path to Power, The, 97, 107, 131
Pecan Bayou, 3
Peck, Gerald W., 31, 41–43
Pedernales Electric Cooperatives, 129–132, 142, 151–152, 200–201, 202, 225
Pedernales River, 3, 114
Pennington, C. R., 64, 71, 80
Perry, E. H., 82
Petri, Bill, 188–193, 196, 197, 198, 206–207, 214
Pickle, J. J. (Jake), 182–183, 210, 222
Piñeda, Álvarez, 5
Plano, Texas, 88
Poage, W. R., 52
Pope, J. F., 19, 20
Possum Kingdom Dam, 110
Potter, Morgan, 144–145
Powell, Ben H., 43, 44
   Ben, Jr., 209
Powell Bend, 219
Powell, Wirtz, Rahut, and Gideon, 150
Pryor, Cactus, 222
Public Works Administration, 37, 45, 47, 53, 56, 60, 65–66, 71–72, 81, 112, 149, 163–164

Q

Quigley, Francis P., 152

R

Radley, Rita Myatt, 234
railroads, 3, 9, 25, 204
Randolph, Beverley, 32, 33, 35, 43, 161
Rauhut, John, 43
Rayburn, Sam, 79, 101, 111, 128, 149

Reconstruction Finance Corporation, 36, 37, 142, 162

Reed, Representative, 63
    S. G., 10

Reinhard, A. J., 65, 71

relief bonds, 62

rice crops, 3, 11–17, 57, 172, 176, 180–181

rice farming, 4, 102–103

Ritter, John R., 124

Rivers and Harbors Act, 8, 149

Rivers and Harbors Committee, 101

Robert Lee Dam, 221

Rockdale Power Project, 226

Roosevelt, Franklin D., 36, 41, 45, 53, 92, 97, 100, 101, 128, 132
    James, 102

rural electrification, 107–112, 113, 126–132, 141–142, 151–157, 190

Rural Electrification Act, 111

Rural Electrification Administration (REA), 126–127, 128–129, 153, 187

S

Saegert, Albert W., 42

San Antonio Public Service Company, 142

San Antonio, Texas, 162

San Bernard River, 6

Sandia, Texas, 136

San Felipe de Austin, 6

San Jacinto Battlefield, 63

San Marcos, Texas, 154

San Saba River, 3

San Saba, Texas, 154

Sansom, R. L., family, 156–157

Savage, ———, 63

Scanlan, John M., 212, 221, 227
    Patricia Wilson, 234

Schmitt, G. E., 185, 192–193

School, Danny, 178

Schulenberg, Texas, 139, 142

Schumann, Merritt, 227, 228

Scott, Will H., 48

Seguin, Texas, 137–138

Selig, Marvin, 212, 227

Senate Bill 178, 169

Seymour, "Muddie," 210
    Sam K., Jr., 186, 192, 195, 198, 210, 222, 234

Shapiro, Harry, 212, 215

Shaws Bend, 227

Shearer, Gordon K., 171

Shelton, Polk, 100, 101

Sheppard, Morris, 38, 41, 116

Shirley Shoals, 27–28

Shivers, Allan, 176, 180

Sims, Bill, 220

Small, Clint, 198, 201, 203, 204, 205

Small, Craig, and Werkenthin, 203

Small, Herring, Craig, and Werkenthin, 198

Smith, E. (Babe), 129–131, 222
    Preston, 196, 198

Soderberg, Elof, 161–162, 196, 197, 200–201, 203, 208, 212–217, 218–219, 220, 222, 226, 227–229
    Mary Virginia, 212–213

soil conservation program, 171–174

South Texas Chamber of Commerce, 138

South Texas Interconnected System, 210

Southern Pacific Railroad, 25

Spires, A. B., 159

Stacy Dam, 219–223

Staple, E. W., 36

Starcke, Max, 135, 136–138, 139–141, 143, 151, 153, 157–158, 159, 162, 166, 170, 174, 180, 185, 186, 194–195

Starcke Dam, 113, 179–180, 194–195, 232

Starcke Park, 137

statewide water plan, 220–223

Stevenson, Coke R., 52, 166, 168, 173, 199

Stinson, ———, 63

Stone, Albert, 52, 120
    Sam, 100, 101

Strahan, J. C., 207

Stranahan, Harris and Co., Inc., 162

Sturrock, J. E., 195

"Sugar Bowl," 10

sugar cane, 10, 12

Sulak, L. J., 118, 124

Sullivan, D. A. J., 113

Summerrow, Ray E., 32, 35, 75

Sutherland, George, 92

Swanson, Bob, 177

Syndicate Power Company, 28, 29, 30,

31, 113

T

Taylor, Thomas U., 19, 24, 120–121
Temple Chamber of Commerce, 110
Tennessee Valley Authority, 128, 229
Texas A&M Extension Service, 172
Texas Brewers Association, 49
Texas Club of New York, 177
Texas Electric Cooperatives Act, 127, 128
Texas League of Municipalities, 138
"Texas, Li'l Darlin' ," 178
Texas Motor Transportation Association, 175
Texas Municipal League, 138
Texas Power and Light Company, 40, 50, 88, 126, 127, 129, 138, 139–142, 152, 154
Texas Public Utility Commission, 225
Texas Railroad Commission, 200, 203, 204
Texas Rangers, 15–16
Texas State Archives, 11
Texas State Board of Water Engineers, 28, 33, 65, 118, 122–123, 124
Texas Supreme Court, 44
Texas Water Commission, 220
"Texland" controversy, 142, 200, 202, 225–226
Thayer, Eugene V. R., 36–37, 56, 58
Thornberry, Homer, 180, 182, 187, 195
Tinstman, Robert M. (Bob), 218
Tipton, Douglas (Red), 177
Tom Miller Dam, 94, 113–114, 138, 151, 233 (*see also* Austin Dam)
Tonkawas, 4
tourist industry, 4
Travis County, 193
Travis County Commissioners Court, 65
Treude, Norman, 16–17

U

U.S. Army Corps of Engineers, 8, 9, 210
U.S. Department of Agriculture, 127, 128
U.S. Department of Interior, Bureau of Reclamation, 70, 72, 81, 87, 92–93, 95, 96–97, 121, 124, 148, 149
U.S. Geological Survey, 2, 124
U.S. Weather Bureau, 123, 147

Umstattd, Mac, 191, 203
University of Texas Board of Regents, 49
utility companies, 14–15, 29, 34, 36, 54, 61, 88–92, 104–106, 109–111, 126–128, 138–142, 163, 174, 184, 187, 190, 193, 211

V

Valero Energy Corporation, 204
Van Zandt, Olan R., 48
Voekel, Aubrey, 198

W

Waelder, Texas, 142
Walker, J. W., 65
Wallace, Doris, 197
Ward, Samuel, 8
water conservation, 180, 181
Waterloo, 18
water mills, 18–19
Water Powers of Texas, The, 19
Waters of the Brazos, A History of the Brazos River Authority 1929–1979, The, 110
watershed reporting system, 123–124
Weddell, Wray, Jr., 194
Weinert, Rudoph A., 43, 120
Werkenthin, Fred, 198, 225
West, Bill, 217
West Texas Chamber of Commerce, 29, 30, 31, 53
West Texas Utilities Company, 29, 30, 33, 34, 37
Western Union Telegraph Company, 24
Whalen, Grover, 144
Wharton County, 14, 15
Wharton, Texas, 15
White, Mark, 220, 221
Whiting, William H. C., 8
Wild, Claud, 101
William P. Carmichael Company, 25–26
Williams, John, 196, 210
Wilson, Patricia (Pinky), 227
Wirtz, Alvin J., 31, 32, 33, 35, 36, 37–38, 39, 40–44, 45, 46, 48–49, 66–67, 68–70, 71, 72–73, 78–79, 96, 98, 99–101, 117–118, 124–125, 135, 139, 150, 151, 174, 180, 181–182
Witt, Edgar, 52
Woodfin, Max, 218

Woodruff, H. C., 52
    Margaret, 195
Woodul, Walter, 96, 103
Woodward, Alva, 165
    Walter, 52, 54, 58, 61
Wooten, Goodall, 119
World's Fair (1939), 143, 144
Worsham, Joe, 139
Wright, Fielding, 176

Wyatt, Oscar, 196, 205

Y

Yarborough, Ralph W., 64, 66, 80, 86,
    187, 189, 222, 234

Z

Zachry, Pat, 206
Zercher, Roger Gilbert, 198